Understanding Poultry Science

Understanding Poultry Science

Editor: Ralph Owens

States Academic Press,
109 South 5th Street,
Brooklyn, NY 11249, USA

Visit us on the World Wide Web at:
www.statesacademicpress.com

© States Academic Press, 2022

This book contains information obtained from authentic and highly regarded sources. Copyright for all individual chapters remain with the respective authors as indicated. All chapters are published with permission under the Creative Commons Attribution License or equivalent. A wide variety of references are listed. Permission and sources are indicated; for detailed attributions, please refer to the permissions page and list of contributors. Reasonable efforts have been made to publish reliable data and information, but the authors, editors and publisher cannot assume any responsibility for the validity of all materials or the consequences of their use.

ISBN: 978-1-63989-543-4 (Hardback)

Trademark Notice: Registered trademark of products or corporate names are used only for explanation and identification without intent to infringe.

Cataloging-in-Publication Data

Understanding poultry science / edited by Ralph Owens.
   p. cm.
Includes bibliographical references and index.
ISBN 978-1-63989-543-4
1. Poultry. 2. Aviculture. 3. Eggs--Production 4. Animal culture. I. Owens, Ralph.
SF487 .U53 2022
636.5--dc23

Table of Contents

Preface VII

Chapter 1 **Genomics Tools for the Characterization of Genetic Adaptation of Low Input Extensively Raised Chickens** 1
Farai Catherine Muchadeyi and Edgar Farai Dzomba

Chapter 2 **Chemical Contaminants in Poultry Meat and Products** 20
Ayhan Filazi, Begum Yurdakok-Dikmen, Ozgur Kuzukiran and Ufuk Tansel Sireli

Chapter 3 **Mycotoxins in Poultry** 40
Ayhan Filazi, Begum Yurdakok-Dikmen, Ozgur Kuzukiran and Ufuk Tansel Sireli

Chapter 4 **Identification of Microbial and Gaseous Contaminants in Poultry Farms and Developing Methods for Contamination Prevention at the Source** 60
Dorota Witkowska and Janina Sowińska

Chapter 5 **How to Control *Campylobacter* in Poultry Farms?: An Overview of the Main Strategies** 82
Michel Federighi

Chapter 6 **Biofilms of *Salmonella* and *Campylobacter* in the Poultry Industry** 100
Daise A. Rossi, Roberta T. Melo, Eliane P. Mendonça and Guilherme P. Monteiro

Chapter 7 **Selenium Requirements and Metabolism in Poultry** 121
Anicke Brandt-Kjelsen, Brit Salbu, Anna Haug and Joanna Szpunar

Chapter 8 **An Overall View of the Regulation of Hepatic Lipid Metabolism in Chicken Revealed by New-Generation Sequencing** 143
Hong Li, Zhuanjian Li and Xiaojun Liu

Chapter 9 **Assessment of Maize (*Zea mays*) as Feed Resource for Poultry** 157
Herbert K. Dei

Chapter 10 **The Effect of Age on Growth Performance and Carcass Quality Parameters in Different Poultry Species** 188
Daria Murawska

Chapter 11 **Poultry Litter Selection, Management and Utilization in the Tropics** 206
Musa I. Waziri and Bilkisu Y. Kaltungo

Permissions

List of Contributors

Index

Preface

In my initial years as a student, I used to run to the library at every possible instance to grab a book and learn something new. Books were my primary source of knowledge and I would not have come such a long way without all that I learnt from them. Thus, when I was approached to edit this book; I became understandably nostalgic. It was an absolute honor to be considered worthy of guiding the current generation as well as those to come. I put all my knowledge and hard work into making this book most beneficial for its readers.

All the birds which are domesticated by humans for the purpose of utilising their fur and eggs are referred to as poultry. The term poultry also includes pigeons known as squabs. Some common examples of poultry are chicken, turkey, geese and ducks. It is eaten across the world since the meat has low fat and high protein content. Poultry are often reared in cages, coops and hatcheries. Poultry science is a study of these birds. It aims to improve the poultry production by producing flavourful meat, healthy eggs and increased resistance to diseases. Poultry are susceptible to infectious agents such as E-coli and salmonella. Poultry science aims to control the outbreak of such diseases. Most of the topics introduced in this book cover new techniques and the applications of poultry science. It strives to provide a fair idea about this discipline and to help develop a better understanding of the latest advances within this field. As this field is emerging at a rapid pace, the contents of this book will help the readers understand the modern concepts and applications of the subject.

I wish to thank my publisher for supporting me at every step. I would also like to thank all the authors who have contributed their researches in this book. I hope this book will be a valuable contribution to the progress of the field.

Editor

Genomics Tools for the Characterization of Genetic Adaptation of Low Input Extensively Raised Chickens

Farai Catherine Muchadeyi and Edgar Farai Dzomba

Abstract

Evolutionary change emanating from differential contribution of genotypes to the next generation can determine success in survival and reproduction in chickens. For extensively raised chickens reared under low-input production systems in smallholder farming areas, conditions of resources deprivation and exposure to diverse and threatening natural selection pressures are common in many countries worldwide. Numerous studies have demonstrated that village chickens and other extensively raised chicken populations represent a valuable source of biodiversity adapted to the local production conditions and selection pressures. Manipulation of their acquired adaptive genetic diversity depends on unravelling the selection footprints in the genomes of these chickens that could point towards candidate genes for traits that enable the animals to survive under the harsh production environments. This chapter summarizes the evidence for chickens' adaptation to extreme environments and describes an inventory of modern tools that could be used in characterizing the production systems of chicken genetic resources. The role of natural selection in shaping the biodiversity of chicken genetic resources is discussed. The continued advancement of biotechnological tools to assess chicken populations has been beneficial to research in genetic adaptation. Genomics tools, as evidenced by assays of whole genome and transcriptome sequences, and single nucleotide polymorphism (SNP) genotypes of chickens, now allow analyses of functional genomic regions that are linked to adaptation. The use of these methods to characterize and investigate signatures of selection in the chicken genomes is highlighted. This chapter looks at how information on the selection hotspots in the chicken genomes can be manipulated to improve genetic adaptation in indigenous chicken populations with the desire to transfer the benefits to other chicken breeds raised under similar production systems.

Keywords: selection footprints, production system, landscape inventory, chicken genetic resources, genetic adaptation

1. Introduction

Chicken production in Africa and many developing countries plays an integral role in smallholder farming systems and relies upon rearing of scavenging indigenous chickens commonly referred to as village chickens [1]. Chickens fulfil several household needs that are cultural, economic and/or social. However, the contribution of indigenous chickens from rural localities to chicken biodiversity has assumed prominence [2]. Purpose-driven artificial selection and breeding for traits of economic importance are undocumented in village chicken populations. Instead, the chickens are regarded as subjects of natural selective pressures applied through factors such as endemic diseases, climate, nutrition and other variables. The result is a variety of plumage colours, productivity levels, body sizes and disease resistance phenotypes. The resultant biodiversity ensures that the chickens can survive in diverse ecological zones by natural selection of genotypes promoting survival fitness. It is therefore hypothesized that the village chicken populations could harbour genomic segments similar to those found in the ancestral chickens *Gallus gallus domesticus*. The gene pool has survived the historical selection pressures and allows village chickens to survive in the harsh and diverse environmental conditions.

In response to the global shift in environmental and market demands for chicken products, the diversity in village and other chicken populations could be needed in future for chickens' genetic advances and improvements. Evolution of chickens and programs for their artificial selection rely on the availability of sufficient levels of genetic variation. Village chickens are regarded as important genetic reservoirs that have evolved in harsh environments over many generations in the absence veterinary and intensive management [3]. Adoption of the free range organic farming systems could result in increased demand and dependency on village chicken genotypes that are already thriving in a very similar production system [4]. Marker-assisted selection and introgression are possible vehicles of harnessing the valuable genetic variation in village chickens for the benefit of high-performing commercial populations [5]. However, the success of genetic improvement strategies relies on a systems approach that ensures that other production constraints are met [5].

This chapter looks at village chicken populations, the characteristics of their production systems and how that shapes their genomic architecture. The chapter describes the production challenges and opportunities experienced in village chicken production systems and how they could be regarded as selection pressures for village chicken populations. The importance of identifying selection footprints in genomes of different village chicken population is highlighted. Lastly, this chapter examines the potential role of next generation sequencing and genotyping technologies aided by advances in statistical genomics in determining signatures of selection and explaining the adaptive genetic diversity found in extensively raised chicken populations.

2. Chicken populations and genotypes

Despite their socio-economic importance and some efforts to characterize them, insufficient knowledge exists about the genetic composition and potential of village chickens in Africa and most developing countries. Their hypothesized value as a genetic reservoir remains uncharacterized [2, 3]. Local chickens across the African continent are commonly referred to as either 'village' or 'indigenous' chickens regardless of their population boundaries. However, the chicken populations consist of phenotypically distinct variants reared by rural farmers in different agro-ecological zones (eco-zones) or farming regions in different parts of Africa. Most countries use the term 'eco-types' to describe chickens from various farming systems characterized by different environmental conditions [6, 7]. A number of molecular studies using autosomal and mitochondrial markers have been conducted with the objective of investigating genetic diversity and structures of village chickens of a number of African countries that include Zimbabwe [8, 9], South Africa [10–12], Tanzania [7], Sudan [13], Kenya and other East African countries [14] and Ethiopia [15].

Recently genome-wide SNP data have been used to further characterize village chicken populations [16–18]. These studies confirmed the notion that local chicken populations of Africa are diverse and could be a genetic reservoir of genes not found in other chicken populations. The previous studies failed to comprehensively characterize the chicken genetic resources in terms of their unique genetic features and adaptation to local conditions. Chicken populations with genetically distinct genotypes often possess rare and peculiar genetic features due to unique alleles and allelic combinations. There should be differentiation of the overall genetic diversity of a population to its adaptive genetic diversity described by the genetic attributes of host chicken populations. This would enable their survival in a certain environments where other populations would fail to thrive. There are challenges in deciphering the genetic causes enabling village chicken populations' survival in extremes of environments. However, identifying the dimension in which chickens have adapted to their local environments could enhance our understanding of, and ability to genetically manipulate, the village chicken genotypes.

3. Characteristics of extensive and smallholder chicken production systems

Village chicken production systems typically have households that keep different poultry species and farm other livestock and crop species under the extensive system of production [1, 19]. Chickens have been found to dominate in number and economic contribution in most village livestock production systems [8, 19, 20]. Flock sizes predominantly range from 4 to 50 birds per household [4, 19, 21–23]. The chickens are mostly non-descript breeds utilized for both meat and egg production [23]. Molecular genetic studies have reported high within-population diversity and a deficiency of population substructures across the chicken populations.

Smallholder farmers have limited resources that are allocated to many enterprises contributing to their livelihoods. As a result, inputs towards chicken production in the form of housing, nutrition and veterinary intervention are minimal and, in many households, are non-existent [19, 24, 25]. The chickens scavenge for their feed [26] and are exposed to the full variability of environmental conditions. Little is understood of how disease epidemiology coupled with poor infrastructure and inadequate diagnostic facilities makes health control challenging [4, 24]. The chickens are raised as mixed flocks combining different age groups. In addition, contacts with wild birds and other livestock species aids in transmission of diseases and jeopardizes biosecurity [1, 27].

Characterization studies of village chicken production systems aim to identify their production challenges and potential. Various studies surveyed and reported the health and disease challenges facing the village chickens from various production systems in Africa and other developing countries [28–32]. Nutrition has been reported as a major challenge due to either insufficient or poor-quality feed [33–35]. Other challenges reported include predation and uncharacterized or unimproved gene pool [36]. Regardless of its shortcomings, the extensive or scavenging method is considered the most important in smallholder chicken production [1, 4]. It is a low-input production system that allows farmers to produce eggs and meat without resorting to expensive poultry feeds, often unaffordable to the rural farmers.

4. Natural selection pressures as forces of genetic adaptation in village chickens

It is evident that the prevalent chicken production constraints present selection pressures on village chicken production systems resulting in certain genes being favored or eliminated from a population. Mortality in village chicken populations is high particularly in young chicks resulting in their failure to reach reproductive maturity and breed, passing their genes to the next generation. Local chickens are raised under harsh, extensive and heterogeneous environments of Southern Africa characterized by inadequate water supply, low quantities of feed which is often of poor quality, and a host of many parasites and disease-causing pathogens [8, 37]. The chickens scavenge for food and water in an environment infested by a number of gastrointestinal parasites and other disease-causing pathogens such that they are continuously re-infected by parasites on different infective stages [38]. The local chickens are regarded as adapted to, or at least tolerant of, the parasite-infested feed resources [39]. The adaptation is considered a valuable attribute that could find application even in the commercial sector under free-range production systems. The mortality due to diseases, parasites, climatic and nutritional stress represents selection pressures whereby the chicken genotypes able to survive the harsh production conditions are able to survive whilst the rest are selected out. Reproductive wastage due to poor hatchability and predation are other selection pressures resulting in fewer breeders contributing to the next generation. To survive this selection pressure, adaption traits such as plumage colour, good mothering ability and the ability to sense danger and respond promptly become important. Dark plumage colour is considered as a camouflage against predators particularly in mature birds [40].

Village chickens face a number of nutritional constraints that include (i) low quality and fluctuating scavenging feed resource base [33, 37, 41]; (ii) inadequate and erratic supplementary feeding that in most cases depends on the availability of crop and household residues at the different households and not so much on the nutritional requirements of the birds; and (iii) presence of anti-nutritional factors in the unprocessed feeds [26]. Chickens are known to compete with humans for grain and other nutritional sources. As a result of human food insecurity in most smallholder farming systems, chickens are often found feeding on residues. In most instances, the composition of such residues as feed does not meet the nutritional requirements of the chickens. Examples of such feeds are maize hauls from mealie-meal processing as well as low quality beans and other legumes residues [26, 33]. Chickens surviving under village chicken production systems should therefore be able to tolerate and adapt to the high roughage diets and anti-nutritional factors found in most of the feeds that the chickens scavenge on. Although protein is not a limiting factor, the worms and insects that the chickens feed on could be vectors of parasites and diseases. It is imperative for the chickens to develop a defense mechanism that prevents or fights infections from the different nutritional sources.

Breeding and selection in village chickens has been largely a natural selection feature and as such the chickens are generally described as non-descript birds not specialized for any particular production traits [9]. However, of concern is that absence of cock-sharing among villages, a lack of pedigree records and disorganized mating structures present problems with inbreeding resulting from breeding too few cocks in flocks where the hens are their close relatives [9, 29]. Although hen to cock ratios of 5:1 has been reported, most farmers in these village chicken production systems have been found to choose a few cocks that they use to propagate into the next generation based on preferred phenotype and morphological features [39]. Farmers have been observed to select based on features such as body size, plumage colour, among other attributes [9]. In many cases, there are no organized breeding programs and these unconventional selection practices could result in inheritance of only a few genes from the few breeding stock of the next generation.

5. Rationale for studying genetic adaptation in local chicken populations

Genetic adaptation is key to survival in harsh production environments where management interventions are limited [4]. Adaptation will enable animals to survive and produce optimally in the face of extreme environmental conditions, limited and poor quality feed and diseases and parasite infections which are characteristics of most village chicken and other livestock production systems. The change in consumer demands resulting in changes in production methods has shifted poultry breeding goals from merely meat and egg production oriented to incorporation of such traits as welfare and disease resistance and other adaptive traits particularly in poultry. It is on this basis that local chickens that have been raised and have adapted to the extensive systems of production are important. With the advent of genomic (GAS) and marker (MAS) assisted selection and marker assisted introgression (MAI) techniques, genes conferring adaptation can be selected to improve resistance or tolerance to selection pressure in a population or can be introgressed into other populations.

Adaptation to local environments is a phenomenon associated with village chickens and other indigenous livestock genetic resources. Village chicken populations could be sources of alleles and allelic combinations developed over long periods of time and conferring adaptive advantages to the chickens carrying them [3]. The drivers for selection in the farming systems of African livestock include socio-cultural factors and potential for livestock survival in wide-ranging environments [42]. To survive, African livestock need to adapt to the varying virulence of communicable diseases, climatic challenges and nutritional constraints [42]. Adaptive and survival traits requisite to chickens raised under typical village chicken production systems for example include the ability to escape predators by running or flying, hens going broody in order to naturally hatch chicks without artificial incubation and birds learning how to scavenge for feed and water resources [33].

Although hypothesized to have developed adaptive mechanisms to survive the harsh environments they are raised in [4, 8], the genetic mechanisms involved in adaptation of village chickens largely remains unclear. The environmental factors which trigger the chickens' adaptation are not known and efforts are still underway to investigate whether that adaptation is genetic and heritable. Questions such as how many genes are involved in adaptation, the nature of the genetic variation conferring the adaptation and whether adaptation utilizes pre-existing genetic variation or new mutations after an environmental challenge remain unanswered.

Knowledge of the adaptive traits developed by local chicken populations is important as part of an inventory of the genetic resources that are present and the useful genes that they might be carrying [42]. Adaptive genetic diversity enables populations' flexibility and survival in changing production environments, consumer demands and environmental and climate changes [4, 43]. A diverse array of genes found in indigenous livestock populations allows populations to survive in diverse and changing environments and should be the target for conservation [43]. It is therefore important that the genetic adaptation is characterized to enable accurate conservation measures.

One challenge in the study and characterization of village chicken populations has been finding appropriate tools and frameworks to characterize adaptive genetic diversity. In the tropics, indigenous chickens genetic resources are plenty and could provide a foundation for adapted breed development through genetic improvement and diversification [44]. Nine major genes of indigenous chickens that could become useful in genetic improvement initiatives have been identified [44]. However, knowledge is still sparse on the genetic constituents of the indigenous chickens of Africa. Efforts to collate information by FAO Domestic Animal Diversity Information System (DAD-IS) points to some unique genetic variants found in the local chicken populations across the African continent. Recent developments in genomics and bioinformatics have opened up platforms to study genetic mechanisms of adaptation as well as ways to manipulate the advantageous alleles and genotypes.

6. Previous attempts to characterize and manipulate adaptive genetic diversity in village chicken populations

Utilization of the vast amounts of genetic variability within village chickens could be useful to their adaptation to local environments through promotion of favourable genetic variants associated with traits such as disease resistance. The unravelling of the genomic architecture for fitness and robustness of the individual chickens is the missing link to the direct application of modern molecular techniques to improve adaptation through breeding programs. Positive selection towards enhanced robustness could advance adaptation of indigenous chickens to harsh and extreme environmental conditions. This would result in chickens that carry alleles necessary for adaptability to the local conditions. The genetic link between the environment and adaptation has not been fully explained but studies abound demonstrating the association between genotypes and the environment for possible inference on genetic adaptation. More data and cases are required to draw clearer conclusions.

Conventional approaches to study adaptive traits in chickens have involved controlled experiments whereby particular environmental challenges are simulated and the response of chickens is monitored [39]. These methods have the limitation that the actual environment where these chicken populations are raised is too heterogeneous in nature and complex to be fully simulated under restricted experimental designs. Resistance to diseases and parasites, for example, are complex traits influenced by both the genes of the animal and the environment under which these animals are living. Under varying environmental conditions, the enteric bacterial pathogens have shown different effects on the intestinal epithelium. Understanding the avian immune system and the response of the chicken as a host is vital in designing prevention strategies for intestinal diseases in poultry. Host response after interactions with pathogens has not yet been fully unravelled.

There has been some work reported using morphological marker genes for genetic improvement in livestock. Crossbreeding experiments of the Fayoumi breed showed an increase in egg production when breeds carrying the naked neck (*Na*) genes were used [45]. Similarly, economically important traits of feed efficiency of chickens under heat stress through use of dwarf gene (*dw*) carrying breeds and egg production and egg weight enhancement by crossing with naked neck (*Na*) and frizzle (*F*) genes-harbouring breeds have been demonstrated [46].

Autosomal microsatellites and mitochondrial DNA sequences have been used to infer on the genetic structures and maternal origins of most village chicken populations of Africa revealing lack of population sub structuring within village chicken populations of a number of countries such as Kenya and other east African countries [14], Zimbabwe [8], Ethiopia [15] and South Africa [12]. The studies have also shown a considerable level of sub structuring between countries. mtDNA studies have shown multiple maternal lineage and high level of genetic diversity in the studied African populations. Core set analysis of the South African chicken populations has shown that they are a reservoir if genetic diversity that could make a significant contribution to the global diversity found in chicken genetic resources ([12]. Other studies have used single nucleotide polymorphism (SNPs) within candidate genes associated with traits of economic importance to study genetic variability and infer on traits such as immune response and growth. Elucidating the genetic variability in indigenous chicken at gene loci important for the immune system *e.g.* Toll like receptor 7 (*TLR7)* nowadays relies on molecular genetic tools. A study used a total of 24 SNPs and 35 haplotypes within the family of Toll Like Receptors genes to investigate genetic variability in immune response traits of Sri Lankan Indigenous chickens [47]. An investigation of the polymorphism at a single *Mx* locus inferred on resistance of populations to Avian Influenza [48]. The genetic diversity of two important immunity candidate genes, the inducible nitric oxide synthase (*INOS*) and natural resistance-associated macrophage protein 1 (*NRAMP1*) in indigenous Malaysian chickens, was carried out using the polymerase chain reaction-restriction fragment polymorphism (PCR-RFLP) [49].

There is currently an increased focus on genetic characterization and conservation of biodiversity globally. Whilst these previous studies have provided the much-needed information on the genetic diversity of chicken genetic resources, there has not been much success in terms of characterizing the adaptive chicken diversity. This has mainly been due to the neutral nature and sparse genome distribution of markers used in earlier studies. Although recent genomic studies using either genomewide SNP or sequence data have highlighted the utility of such genomic tools in characterizing adaptive genetic diversity [16–18] very few studies have actually used the data to infer adaptive genetic diversity in chicken populations [18].

7. Genomics tools for characterization of signatures of selection and genetic adaptation

There have been significant developments in our understanding of the genome of chickens and the statistical applications that can be used in elucidation of genetic adaptation and robustness of local livestock populations including chickens. The chicken genome was sequenced in 2004, which opened new avenues for the exploration the genomic architecture and advance genomic studies. Alternative genomic variants such as Linkage disequilibrium (LD) blocks, Copy Number variations (CNVs) and Runs of Homozygosity (ROH) have increasingly found use in studying genetic diversity and drawing of inferences on genetic adaptation in chicken populations. Contrary to prior markers that were either found in non-coding regions of the genome or sparsely distributed, genomics tools present the required genomic coverage to investigate regions associated with traits of economic importance.

7.1. Copy number variations

Copy number variations (CNVs) have gained enormous research recognition in recent times as genomic structural variants of interest in chicken biodiversity studies [50, 51]. Copy number variations (CNVs) are defined as deletions, duplications or insertions, which change the genomic diploid state of an individual [52–54]. The last few years have seen major advances in genomics demonstrating that the alterations in the DNA could have no effect phenotypically, or account for traits for adaptation and even underlie disease etiology and susceptibility [55, 56] thereby providing increasing evidence of their contribution to breed diversity. Prevalence of CNVs have been reported in chickens and associations suggested with traits of economic importance [50, 57]. Bioinformatics and statistical pipelines have been developed to screen for CNVs and infer on biological functions [56, 57] and associated CVNs to resistance to Marek disease in divergent chicken lines. CNVs that have potential roles in economically important traits in chickens have been suggested [58]. Other studies also associated CNVs to diseases [59].

Numerous reports produced from CNV studies were mostly restricted to commercial or experimental chicken lines and, thus far to the best of our knowledge, none were used in village or extensively raised chicken populations. However, there is scope to use CNVs in these nondescript and uncharacterized chicken populations that have been exposed to natural selection pressures.

7.2. Runs of homozygosity

Runs of homozygosity (ROH) are lengthy continuous regions of the genome where, in the diploid state, the copies inherited from the parents are identical. These stretches have been found to estimate inbreeding coefficients that are much better at detecting the overall burden of rare and recessive mutations based on high-throughput, beadchip single nucleotide polymorphism (SNP) genotypes [60]. Associations have been made between ROH and genetic defects arising from intensive selection and inbreeding in other species [61]. There have been very few studies using runs of homozygosity in village chickens. Previous studies have suggested inbreeding to be a problem in village chickens that are raised as small fragmented flocks with a lot of cock sharing of few male animals within villages [16]. Runs of homozygosity have potential as estimators of inbreeding through identification of genomic regions they affect [62].

7.3. Linkage disequilibrium and haplotype blocks

Linkage disequilibrium (LD) is defined as non-random association of alleles more than one loci [63, 64]. In genetics, LD is a useful tool increasingly finding application in the genomic era with the availability of data from high-density SNP panels. The decay and extent of LD at a pair-wise distance can be used to estimate trends in effective population size [17, 65, 66]. LD will therefore be useful in extensively raised chicken populations in smallholder farming systems where, due to non-collection of pedigree information, calculation of population genetic parameters could be possible. LD estimates have been used to determine trends in effective population size in egg and meat producing chicken lines [64]. A study used the chicken 60K panel to determine LD and estimate effective population sizes in extensively raised chicken populations of southern Africa [16]. Estimates of inbreeding and trends in effective population sizes provide information required to monitor and conserve village chicken populations in the absence of pedigree data. Such information also sheds light on demographic forces that influence genetic diversity of village chicken populations.

7.4. Landscape genomics

Landscape genomics combines genome-wide information with geo-environmental resources analysis to identify potentially valuable genetic material [67, 68]. With regards to village chickens, the aim for landscape genomics is to learn from their co-evolution and production systems and use the knowledge gained to better match different genotypes with production circumstances. Regions where chicken genomes face selection pressure from environmental conditions such as high temperatures, lack of water in drought *etc.* are expected to show higher genomic divergence across habitats compared to the neutral genome background. By performing genomewide-scans on chickens from populations living in different habitats or across agro-ecological cline, from dry to wet areas, for example, landscape genomics offers the opportunity of analyzing the immune response of chickens across a naturally occurring environmental gradient [67]). By collecting samples from chickens from different environmental conditions and assessing their phenotypic performance and genotypes at certain loci, one can have an idea of how the environment has influenced the immune system of the chickens to infection by pathogens. In previous chicken studies, it has been shown that agro-ecological zones or climatic regions have an important role in defining free ranging chicken production systems [8]. The eco-zones influence the type of feed animals has as well as disease and parasite pathogen profiles. Agro-ecological zone also influences the physiological function and response of chickens to disease-causing pathogens thereby determining immune response status. As in most species, chickens raised within a certain locality tend to develop adaptive traits that help them survive in the given environmental challenges. By applying landscape genomic techniques, one will be able to overlay the genetic profiles of chickens over the environmental coordinates, identify and characterize genes conferring adaptive features to the different environmental conditions. As a new tool, reports of case studies on the application of landscape genomics are sparse for village chickens. The Mediterranean goat breeds have been characterized using landscape genomic approaches and results in one study showed regions of genomes being under different selection pressures in different environments [67]. One of the major requirements for landscape genomics is a dense set of molecular markers both neutral and those under selection to be able to make inferences on the role of selection versus neutral evolutionary forces on shaping genetic diversity. Both next generation sequencing and high-density SNP panels have provided such platforms.

7.5. Signatures of selection

An alternative approach to ascertain regions of genome with relevant genes for genetic adaptation and robustness in chickens could be the detection of selection signatures using genomewide SNP data or sequences [69, 70]. The principle that loci across the genome are influenced by genome-wide evolutionary forces like migration, random genetic drift, mutation etc is the basis for population genomics. Locus-specific forces *e.g.* selection help create particular patterns of variability in the regions of associated genes And loci [70]. Comparison of the genetic diversity of across numerous loci on the genome, it is possible to uncover loci exhibiting an atypical variation pattern which are could be linked to areas in the genome affected by selection [69]. Signatures of selection have been investigated in a number of chicken populations [17, 18]. Signature of selection analysis does not necessarily require conventionally-measured performance as in GWAS analysis. This makes it a suitable tool to screen for advantageous loci in extensively raised village chicken populations.

7.6. Comparative genomics

With plummeting costs of whole genome sequencing, more and more sequence data are being generated across different species. Comparative genomics involves sequencing and comparing genomes of different species in search of genomic regions differentiating species. In several avian species including chickens, use of comparative analysis of genome sequences has emerged as a powerful tool for understanding metabolic processes and gene functions influenced by selection [71]. Inter-species comparison of selection signatures can play a role in detecting relevant functional polymorphisms in experiments using populations collected from a variety of environmental conditions. An analysis comparing the evolution of chicken (*Gallus gallus*) and zebra finch (*Taeniopygia guttata*) genomes [71] showed that an important cytokine receptor, the interleukin 4 receptor alpha-chain (*IL-4Rα*), has significantly an abundance of substitutions at nonsynonymous sites which could be a crucial mechanism in its adaptive evolution.

7.7. Transcriptomic analysis

Complete sequencing of the chicken genome and the availability of data on various transcriptomes of specific tissues ushered in another field of functional genomic analyses of chickens. A complete set of transcribed mRNA and the related amounts for specific stages of development or physiology is known as a transcriptome [51]. Studying transcriptomes is now considered essential for unravelling the functional elements of the genome and understanding the molecular and physiological processes in cells and tissues in chickens. Transcriptome analysis identifies genes expressed and gene pathways affected in chickens raised under the heterogenous village chicken production systems, giving an indication of potential candidate genes for adaptation and robustness. Transcriptome sequencing and gene expression analysis were used to investigate genes conferring resistance to colonization of chicken intestines by *Campylobacter jejuni* [72]. A study of gene expression profiles reported differentially expressed genes in chicken intestines infected with *Ascaridia galli* [73]. With the chicken genome and other functional genomic tools available, generated transcripts can easily be aligned to the chicken genome and other transcripts of reference chickens to assess the gene expression profile under different conditions.

8. Genomic tools in village chicken breed improvement programs

Selection and breeding are among the key forces moulding village chickens' genomic architecture and regulating their genetic diversity. Although a majority of chickens found in most smallholder production systems are uncharacterized and regarded as unimproved, farmers have indicated preferences for certain phenotypes that they select for in hens and cocks for breeding [9, 74]. However, these traditional breeding activities are conducted in the absence of pedigree records, without collecting performance data and lacking institutional support thereby creating a false impression that no selection and breeding programs for village chicken populations are available. Before the advent of genomics, it has thus been difficult to assess the impact of any selection and breeding in village chickens.

The nature of village chicken production makes selection and breeding challenging. The chickens are raised communally with a lot of sharing of breeding stock between households within communities. Mating is indiscriminate and left to chances of breeding stock mixing during scavenging in the communal spaces. Due to limited resources and extension support, farmers tend to keep breeding stock for longer periods resulting in overlapping generations. In other systems, farmers have access to exotic breeds adding another dimension to genetic improvement through crossbreeding. However, all these factors present challenges for having organized breeding strategies for village chickens of setting up breed improvement programs for village chickens.

Genomics can provide valuable information to facilitate selection of animals in breeding programs. Genomics can be used to reconcile pedigree and define breeding population boundaries. Genomics, bioinformatics and statistics are also finding application in identifying loci and regions in the genome that are adapted to specific environments. When breed improvement programs are implemented, genomics can be used to monitor genetic progress and the direction thereof.

Community based breeding programs (CBBPs) refer to approaches of improving livestock genetics through incorporating farmer participation into selection and breeding processes, from inception through to implementation. Under circumstances where livestock keepers already rear mixed flocks of livestock together such as in communal grazing systems, the implementation of community-based breeding programs is appropriate [75, 76]. This would be beneficial through shared production capital (animal genotypes, grazing land etc.) and pooling of resources and services (veterinary, feeding and marketing) thereby enabling joint processes of making decisions. Key to community based breeding strategies is the consideration of farmers' views, needs and decisions, and to encourage active participation throughout the life-cycle of the program.

Genomics plays a crucial role in complementing CBBPs through provision of information that would otherwise be out of reach in the marginalized and low-input production systems. Genomics also provides information on the genetic architecture of populations, which is a requisite step in setting up selection programs. Genomics overcomes the challenge of missing pedigree and breed relations information which is a characteristic of most communal production systems where records are not kept as is seen in smallholder village chicken production systems. Numerous case studies exist of CBBPs combining with genomics technologies successfully. CBBPs have proven to be highly successful in Kenya, Bolivia, Ethiopia, Mexico and Peru. In these case studies, the productivity and profitability of indigenous breeds increased in the CBBPs without undermining the resilience and genetic integrity of the populations and potentially reducing genetic diversity. Although no reports of CBBPs are available for chickens to our knowledge, there is potential for combining genomics and CBBPs for the improvement of village chicken populations predominantly raised extensively as large interbreeding flocks in shared environments.

9. Conclusion

Village chicken populations are a source of unique and adapted genotypes that are robust enough to survive in harsh environmental conditions with minimum human interventions.

High throughput next generation sequencing and genotyping tools coupled with advances in statistical genomics are now allowing a comprehensive assessment of the adaptive genetic diversity in livestock species. Analyses of transcriptomes and whole genome sequences and SNP genotypes are tools that can be used to reveal the genetic composition of village chicken populations. Studies have opened doors to analyses of CNVs, ROH and use other bioinformatics and statistical tools such as linkage disequilibrium, comparative genomic analysis and transcriptome profiling to get a deeper insight on evolution of village chicken genomes to adapt to their environments.

Selection and breeding of village chickens is challenged by absence of pedigree and performance data to identify superior animals for use in the next generation. Genomic tools could be the panacea allowing selecting and breeding of chickens outside conventional breed improvement frameworks. In combination with strategies such as CBBPs, there are potential benefits of integrating genomics in the characterization and utilization of the genetic diversity the village chicken populations embody.

Author details

Farai Catherine Muchadeyi[1*] and Edgar Farai Dzomba[2]

*Address all correspondence to: muchadeyif@arc.agric.za

1 Agricultural Research Council, Biotechnology Platform, Pretoria, South Africa

2 Discipline of Genetics, School of Life Sciences, University of KwaZulu-Natal, Pietermaritzburg, South Africa

References

[1] Kitalyi AJ. Village chicken production systems in rural Africa: Household food security and gender issues. Food & Agriculture Org. Rome, Italy; 1998.

[2] Delany ME. Genetic variants for chick biology research: from breeds to mutants. Mechanisms of Development. 2004;121(9):1169-77.

[3] Hall SJ, Bradley DG. Conserving livestock breed biodiversity. Trends in Ecology and Evolution. 1995;10(7):267-70.

[4] Aboe PAT, Boa-Amponsem K, Okantah SA, Butler EA, Dorward PT, Bryant MJ. Free-range village chickens on the Accra plains, Ghana: Their husbandry and productivity. Tropical Animal Health and Production. 2006;38(3):235-48.

[5] Marshall K, Quiros-Campos C, van der Werf JHJ, Kinghorn B. Marker-based selection within smallholder production systems in developing countries. Livestock Science. 2011;136(1):45-54.

[6] Tadelle D, Kijora C, Peters K. Indigenous chicken ecotypes in Ethiopia: growth and feed utilization potentials. International Journal of Poultry Science. 2003;2(2):144-52.

[7] Msoffe PLM, Mtambo MMA, Minga UM, Juul-Madsen HR, Gwakisa PS. Genetic structure among the local chicken ecotypes of Tanzania based on microsatellite DNA typing. African Journal of Biotechnology. 2005;4(8):768-71.

[8] Muchadeyi FC, Eding H, Wollny CBA, Groeneveld E, Makuza SM, Shamseldin R, et al. Absence of population substructuring in Zimbabwe chicken ecotypes inferred using microsatellite analysis. Animal Genetics. 2007;38(4):332-9.

[9] Muchadeyi FC, Wollny CBA, Eding H, Weigend S, Simianer H. Choice of breeding stock, preference of production traits and culling criteria of village chickens among Zimbabwe agro-ecological zones. Tropical Animal Health and Production. 2009;41(3): 403-12.

[10] Van Marle-Koster E, Nel L. Genetic markers and their application in livestock breeding in South Africa: a review: review article. South African Journal of Animal Science. 2003;33(1):1-10.

[11] Mtileni B, Muchadeyi F, Maiwashe A, Groeneveld E, Groeneveld L, Dzama K, et al. Genetic diversity and conservation of South African indigenous chicken populations. Journal of Animal Breeding and Genetics. 2011;128(3):209-18.

[12] Mtileni B, Muchadeyi F, Maiwashe A, Chimonyo M, Groeneveld E, Weigend S, et al. Diversity and origin of South African chickens. Poultry Science. 2011;90(10):2189-94.

[13] Wani CE, Yousif IA, Ibrahim ME, Musa HH. Molecular characterization of Sudanese and Southern Sudanese chicken breeds using mtDNA D-Loop. Genetics Research International. 2014;2014.

[14] Mwacharo JM, Bjornstad G, Mobegi V, Nomura K, Hanada H, Amano T, et al. Mitochondrial DNA reveals multiple introductions of domestic chicken in East Africa. Molecular Phylogenetics and Evolution. 2011;58(2):374-82.

[15] Goraga Z, Weigend S, Brockmann G. Genetic diversity and population structure of five Ethiopian chicken ecotypes. Animal Genetics. 2012;43(4):454-7.

[16] Khanyile KS, Dzomba EF, Muchadeyi FC. Population genetic structure, linkage disequilibrium and effective population size of conserved and extensively raised village chicken populations of Southern Africa. Frontiers in Genetics. 2015;6:13.

[17] Wragg D, Mwacharo JM, Alcalde JA, Hocking PM, Hanotte O. Analysis of genome-wide structure, diversity and fine mapping of Mendelian traits in traditional and village chickens. Heredity (Edinb). 2012;109(1):6-18.

[18] Fleming DS, Koltes JE, Markey AD, Schmidt CJ, Ashwell CM, Rothschild MF, et al. Genomic analysis of Ugandan and Rwandan chicken ecotypes using a 600 k genotyping array. BMC Genomics. 2016;17(1):1.

[19] Sonaiya E, editor The context and prospects for development of smallholder rural poultry production in Africa. CTA-Seminar Proceedings on Smallholder Rural Poultry Production; 1990.

[20] Gueye EF. Employment and income generation through family poultry in low-income food-deficit countries. World Poultry Science Journal. 2002;58(4):541-57.

[21] Muchadeyi FC, Sibanda S, Kusina NT, Kusina JF, Makuza SM. Village chicken flock dynamics and the contribution of chickens to household livelihoods in a smallholder farming area in Zimbabwe. Tropical Animal Health and Production. 2005;37(4):333-44.

[22] Abdelqader A, Wollny C, Gauly M. Characterization of local chicken production systems and their potential under different levels of management practice in Jordan. Tropical Animal Health and Production. 2007;39(3):155-64.

[23] Mtileni B, Muchadeyi F, Maiwashe A, Phitsane P, Halimani T, Chimonyo M, et al. Characterisation of production systems for indigenous chicken genetic resources of South Africa. Applied Animal Husbandry & Rural Development. 2009;2:18-22.

[24] Chitate F, Guta M, editors. Country report: Zimbabwe. SADC Planning Workshop on Newcastle Disease Control in Village Chickens, Canberra, Australia; 2001.

[25] Mack S, Hoffmann D, Otte J. The contribution of poultry to rural development. World's Poultry Science Journal. 2005;61(01):7-14.

[26] Gunaratne SP, Chandrasiri AD, Hemalatha WA, Roberts JA. Feed resource base for scavenging village chickens in Sri Lanka. Tropical Animal Health and Production. 1993;25(4):249-57.

[27] Onapa MO, Christensen H, Mukiibi G, Bisgaard M. A preliminary study of the role of ducks in the transmission of Newcastle disease virus to in-contact rural free-range chickens. Tropical Animal Health and Production. 2006;38(4):285-9.

[28] Mapiye C, Mwale M, Mupangwa JF, Chimonyo M, Foti R, Mutenje MJ. A research review of village chicken production constraints and opportunities in Zimbabwe. Asian-Australian Journal of Animal Sciences. 2008;21(11):1680-8.

[29] Gondwe TN, Wollny CB. Local chicken production system in Malawi: household flock structure, dynamics, management and health. Tropical Animal Health and Production. 2007;39(2):103-13.

[30] Moreki J, Dikeme R, Poroga B. The role of village poultry in food security and HIV/AIDS mitigation in Chobe District of Botswana. Livestock Research for Rural Development. 2010;22(3):1-7.

[31] Okeno TO, Kahi AK, Peters KJ. Characterization of indigenous chicken production systems in Kenya. Tropical Animal Health and Production. 2012;44(3):601-8.

[32] Malatji DP, Tsotetsi AM, van Marle-Koster E, Muchadeyi FC. A description of village chicken production systems and prevalence of gastrointestinal parasites: Case studies in Limpopo and KwaZulu-Natal provinces of South Africa. Onderstepoort Journal of Veterinary Research. 2016;83(1):a968.

[33] Maphosa T, Kusina J, Kusina N, Makuza S, Sibanda S. A monitoring study comparing production of village chickens between communal (Nharira) and small-scale commercial (Lancashire) farming areas in Zimbabwe. Livestock Research for Rural Development. 2004;16(7):2004.

[34] Kebede H, Hussen DO, Melaku A. Study on status and constraints of village poultry production in Metema District, North-Western Ethiopia. American-Eurasian Journal of Scientific Research. 2012;7(6):246-51.

[35] Blackie S. Village chicken production system in the greater Accra Region Ghana. Journal of Biology, Agriculture and Healthcare. 2014;4.

[36] Khobondo J, Muasya T, Miyumo S, Okeno T, Wasike C, Mwakubambanya R, et al. Genetic and nutrition development of indigenous chicken in Africa. Livestock Research for Rural Development. 2015;27:7.

[37] Mtileni B, Muchadeyi F, Weigend S, Maiwashe A, Groeneveld E, Groeneveld L, et al. A comparison of genetic diversity between South African conserved and field chicken populations using microsatellite markers. South African Journal of Animal Science. 2010;40(5):462-6.

[38] Permin A, Esmann JB, Hoj CH, Hove T, Mukaratirwa S. Ecto-, endo- and haemoparasites in free-range chickens in the Goromonzi District in Zimbabwe. Preventive Veterinary Medicine. 2002;54(3):213-24.

[39] Abdelqader A, Gauly M, Wollny CBA, Abo-Shehada MN. Prevalence and burden of gastrointestinal helminthes among local chickens, in northern Jordan. Preventive Veterinary Medicine. 2008;85(1-2):17-22.

[40] Pedersen CV. Production of semi-scavenging chickens in Zimbabwe. PhD Thesis The Royal Veterinary and Agricultural University, Copenhagen, Denmark. 2002.

[41] Mwalusanya N, Katule A, Mutayoba S, Mtambo M, Olsen J, Minga U. Productivity of local chickens under village management conditions. Tropical Animal Health and Production. 2002;34(5):405-16.

[42] Hanotte O, Dessie T, Kemp S. Time to tap Africa's livestock genomes. Science. 2010;328(5986):1640-1.

[43] Hoffmann I. Climate change and the characterization, breeding and conservation of animal genetic resources. Animal Genetics. 2010;41:32-46.

[44] Horst P. Native fowl as reservoir for genomes and major genes with direct and indirect effects on the adaptability and their potential for tropically orientated breeding plans. Archiv fuer Gefluegelkunde (Germany, FR). 1989. 53: 93-98.

[45] Mathur P, El Hammady H, Sharara H, editors. Specific use of high yielding strains carrying major genes for improving performance of local fowls in the tropics (Case Study: Upper Egypt). Proceedings DLG Symposium on Poultry Production in Developing Countries, June; 1989.

[46] Mathur P, Horst P. Temperature stress and tropical location as factors for genotype×environment interactions. Colloques de l'INRA (France). 1989.

[47] Bulumulla P, Silva P, Jianlin H. Genetic diversity at Toll like receptor 7 (TLR7) gene of Sri Lankan indigenous chicken and Ceylon jungle fowl (Gallus lafayetti). Tropical Agricultural Research. 2011;22(4) 367-373.

[48] Ommeh S, Jin L, Eding H, Muchadeyi F, Sulandari S, Zein M, et al. Geographic and breed distribution patterns of an A/G polymorphism resent in the Mx gene suggests balanced selection in village chickens. International Journal of Poultry Science. 2010;9:32-8.

[49] Tohidi R, Idris I, Panandam JM, Hair-Bejo M. Analysis of genetic variation of inducible nitric oxide synthase and natural resistance-associated macrophage protein 1 loci in Malaysian native chickens. African Journal of Biotechnology. 2011;10(8):1285-9.

[50] Han R, Yang P, Tian Y, Wang D, Zhang Z, Wang L, et al. Identification and functional characterization of copy number variations in diverse chicken breeds. BMC Genomics. 2014;15(1):1-10.

[51] Wang X, Byers S. Copy number variation in Chickens: a review and future prospects. Microarrays. 2014;3(1):24-38.

[52] Crooijmans RP, Fife MS, Fitzgerald TW, Strickland S, Cheng HH, Kaiser P, et al. Large scale variation in DNA copy number in chicken breeds. BMC Genomics. 2013;14(1):1.

[53] McCarroll SA, Kuruvilla FG, Korn JM, Cawley S, Nemesh J, Wysoker A, et al. Integrated detection and population-genetic analysis of SNPs and copy number variation. Nature Genetics. 2008;40(10):1166-74.

[54] Alkan C, Coe BP, Eichler EE. Genome structural variation discovery and genotyping. Nature Reviews Genetics. 2011;12(5):363-76.

[55] Henrichsen CN, Chaignat E, Reymond A. Copy number variants, diseases and gene expression. Human Molecular Genetics. 2009;18(R1):R1-R8.

[56] Liu GE, Bickhart DM. Copy number variation in the cattle genome. Functional & Integrative Genomics. 2012;12(4):609-24.

[57] Yan YY, Yang N, Cheng HH, Song JZ, Qu LJ. Genome-wide identification of copy number variations between two chicken lines that differ in genetic resistance to Marek's disease. BMC Genomics. 2015;16(1):1.

[58] Jia P, Jin H, Meador CB, Xia J, Ohashi K, Liu L, et al. Next-generation sequencing of paired tyrosine kinase inhibitor-sensitive and -resistant EGFR mutant lung cancer cell lines identifies spectrum of DNA changes associated with drug resistance. Genome Research. 2013;23(9):1434-45.

[59] Abernathy J, Li X, Jia X, Chou W, Lamont SJ, Crooijmans R, et al. Copy number variation in Fayoumi and Leghorn chickens analyzed using array comparative genomic hybridization. Animal Genetics. 2014;45(3):400-11.

[60] Ku CS, Naidoo N, Teo SM, Pawitan Y. Regions of homozygosity and their impact on complex diseases and traits. Human Genetics. 2011;129(1):1-15.

[61] Bosse M, Megens H-J, Madsen O, Paudel Y, Frantz LAF, Schook LB, et al. Regions of homozygosity in the porcine genome: consequence of demography and the recombination landscape. PLoS Genetics. 2012;8(11):e1003100.

[62] Kim ES, Sonstegard TS, Van Tassell CP, Wiggans G, Rothschild MF. the relationship between runs of homozygosity and inbreeding in Jersey cattle under selection. Plos One. 2015;10(7):e0129967.

[63] Hedrick PW Genetics of Populations. Jones & Bartlett Learning. Sudbury, Massachusetts, USA; 2011.

[64] Fu W, Dekkers JC, Lee WR, Abasht B. Linkage disequilibrium in crossbred and pure line chickens. Genetics Selection Evolution. 2015;47(1):1.

[65] Andreescu C, Avendano S, Brown SR, Hassen A, Lamont SJ, Dekkers JC. Linkage disequilibrium in related breeding lines of chickens. Genetics. 2007;177(4):2161-9.

[66] Lu JT, Wang Y, Gibbs RA, Yu F. Characterizing linkage disequilibrium and evaluating imputation power of human genomic insertion-deletion polymorphisms. Genome Biology. 2012;13(2):1.

[67] Joost S, Bonin A, Bruford MW, Després L, Conord C, Erhardt G, et al. A spatial analysis method (SAM) to detect candidate loci for selection: towards a landscape genomics approach to adaptation. Molecular Ecology. 2007;16(18):3955-69.

[68] Schwartz MK, McKelvey KS, Cushman SA, Luikart G. Landscape genomics: a brief 28 perspective. Spatial complexity, informatics, and wildlife conservation: Springer Tokyo Japan; 2010.

[69] Rubin CJ, Zody MC, Eriksson J, Meadows JR, Sherwood E, Webster MT, et al. Whole-genome resequencing reveals loci under selection during chicken domestication. Nature. 2010;464(7288):587-91.

[70] Elferink MG, Megens H-J, Vereijken A, Hu X, Crooijmans RP, Groenen MA. Signatures of selection in the genomes of commercial and non-commercial chicken breeds. PLoS One. 2012;7(2) 1-11.

[71] Downing T, Lynn DJ, Connell S, Lloyd AT, Bhuiyan A, Silva P, et al. Evidence of balanced diversity at the chicken interleukin 4 receptor alpha chain locus. BMC Evolutionary Biology. 2009;9(1):1.

[72] Li X, Swaggerty C, Kogut M, Chiang H, Wang Y, Genovese K, et al. Caecal transcriptome analysis of colonized and non-colonized chickens within two genetic lines that differ in caecal colonization by Campylobacter jejuni. Animal Genetics. 2011;42(5):491-500.

[73] Malatji D, Tsotetsi A, van Marle-Koster E, Muchadeyi F. Population genetic structure of Ascaridia galli of extensively raised chickens of South Africa. Veterinary Parasitology. 2016;216:89-92.

[74] Dana N, van der Waaij LH, Dessie T, van Arendonk JA. Production objectives and trait preferences of village poultry producers of Ethiopia: implications for designing breeding schemes utilizing indigenous chicken genetic resources. Tropical Animal Health and Production. 2010;42(7):1519-29.

[75] Gizaw S, Komen H, van Arendonk JA. Participatory definition of breeding objectives and selection indexes for sheep breeding in traditional systems. Livestock Science. 2010;128(1):67-74.

[76] Haile A, Mirkena T, Duguma G, Tibbo M, Okeyo M, Rischkowsky B, et al. Community based sheep breeding programs: tapping into indigenous knowledge. Livestock Research for Rural Development. 2013;25:219.

Chemical Contaminants in Poultry Meat and Products

Ayhan Filazi, Begum Yurdakok-Dikmen,
Ozgur Kuzukiran and Ufuk Tansel Sireli

Abstract

Consumption of poultry meat and products has increased as a consequence of economic crisis, driven by several factors, while people keep away from high priced beef/lamb meat or meat products. Meanwhile, due to this increasing demand in industry resulting strict measures in disease control and environmental factors, these products may involve some chemical and natural compounds with hazardous properties at detectable or even very low concentrations. Among these compounds, residues are of concern, including veterinary drugs, environmental pollutants (such as dioxins, pesticides, and phthalates), natural contaminants (mycotoxins, etc), and/or phytosanitary substances accidentally contaminating poultry product during production or marketing stages. In order to keep the consumers safe from the harmful/undesirable effects due to these compounds, such as genotoxic, immunotoxic, carcinogenic, teratogenic, or endocrine disrupting effects, new strategies and concepts for poultry food security have been emerged and developed globally. This chapter includes detailed information on the residues of some potential chemical contaminants in poultry meat and products (eggs, etc.) along with risk analysis regarding their hazardous effects and detection in various matrices.

Keywords: contaminants, egg, meat, poultry, residues

1. Introduction

Air, water, soil, and food are vital constituents of the human environment. While these sources directly affect the quality of human life, the risk of contamination with various pollutants in this industrialized new era is unfortunately inevitable [1]. As the main source of nutrients, food itself contains chemical and natural compounds with hazardous properties. Among these hazardous contaminants, the most important ones are the chemical residues including veterinary drugs, pesticides, and dioxins. With the increased awareness of health and increasing demand on health food, food security standards were developed for the protection of consumers for their adverse health effects. In terms of a wide definition, food safety is defined as a multidisciplinary field including the production, preparation, and the conservation of food in order to protect consumer safety for foodborne diseases in accordance and guidance with the related legislations. Food security, as an evolving scientific field, is an ongoing process, starting with the production until the consumption of the final product [2].

In recent years, consumption of poultry meat and products has increased as a consequence of economic crisis, as people avoid high-priced food. Although egg consumption diminished remarkable in the last period of the twentieth century due to its cholesterol content, recent advances in research provides more evidence on the positive health effects increasing the consumption trades [3].

Residues are substances that can occur in food and feedstuffs naturally or anthropogenically as the accidental, intentional (adulteration), or environmental (persistent organic pollutants like dioxins) contamination of the food with veterinary drugs or phytosanitary products during the production or marketing stages [4]. Codex Alimentarius defines "contaminant" as "any substance not intentionally added to food, which is present in such food as a result of the production, manufacture, processing, preparation, treatment, packaging, transport, or holding of such food or as a result of environmental contamination. The term does not include insect fragments, rodent hairs, and other extraneous matter" [5]. Therefore, all food products are at risk of contamination from several resources, and poultry meat and products are no exemptions. Drug residues can be prevented by not using the substance in animal production, where legal monitoring procedures are applied; meanwhile, these contaminants can be difficult to exempt completely due to the background level of pollution in the environment [4, 6].

The development of a primary production and processing standard for poultry meat uses an approach that investigate the sources of potential chemical risks, which may be introduced at different points through the primary production and processing chain. Poultry meat and products supply chain is divided into four distinct steps: primary production, processing, retail, and consumer. At each of these steps, poultry meat and products may be directly or indirectly exposed to chemicals [3]. Direct exposure results when a compound is present in raw food materials, whereas in indirect exposure, contaminants cross into food during processing, storage, packaging, or preparation. Indirect contaminants also include substances that become toxic and harmful to people due to food-processing practices. Indirect pollution is most frequently the result of unawareness, lack of education of food handlers, insufficient and awful places, or inappropriate handling applications [7].

In all production stage, there is a risk of iatrogenic contamination of multiple substances such as antimicrobials. These substances are essential for poultry production because of their effects in support of health, welfare, and performance, as well as the reduction of manufacturing costs and final costs for the consumer. Additionally, they are substantial to decrease spread of potentially pathogenic organisms from animals to humans and to the environment [8]. Several substances among these drugs used in the treatment have the potential to constitute residues in edible tissues and other food products, which can potentially cause adverse health effects including allergy, pharmacological/toxicological effects, antimicrobial resistance/reduced susceptibility, change of gut microflora, and endocrine disruption in consumers. Therefore, it is substantial that manufacturers, veterinarians, and all other professionals included in food production are aware of the residues and abide by the regulations and the instructions regarding the prudent use, for protecting consumers from potentially detrimental levels of residues [3].

Although most consumers are mainly worried about the residues of veterinary drugs in their food, there are many more potential contaminants in the environment, which are more likely to contaminate the product from various resources. These include phthalates, persistent organic pollutants, various emerging toxic elements, and pesticides [2]. Some of these compounds in residual amounts in poultry meat and eggs have important deleterious effects and known to have genotoxic, immunotoxic, carcinogenic, teratogenic, or endocrine disrupting effects [4].

Contaminants and residues require a comprehensive health apprehension, leading to strict regulations of their concentrations in products decided by national and internationally authorities. For this reason, analysis of relevant chemical substances is a major part of food safety programs to provide consumer safety and agreement with regulatory limits. Modern testing procedures can identify known chemical substances in complex food matrices at very low levels. Additionally, they can also help to reveal and determine new or unexpected emerging chemical substances [9]. This chapter summarizes some potential chemical contaminants and residues in poultry meat and eggs, their hazards, and analysis.

2. Veterinary drugs

Veterinary drugs are generally prescribed for the treatment and prevention of the diseases compromising mostly coccidia, ectoparasites, fungi, and bacterial infections in poultry. Nowadays, a large part of drug use in poultry farming is prophylactic, with the bulk of medications including mainly anticoccidial substances and antibacterial growth promoters [10]. Poultry production systems are closely related to animal and human health; which also has a direct effect on the environment. Therefore, critical risk assessment in the overall production and processing system has great importance [11].

Poultry feed is typically composed of corn and soybean meal mixtures, including several vitamins and minerals, and generally contains two or three medications; which comprises 68% of total production costs. For each development phase, the amounts of the content differ, where starter, grower, finisher, and layer feeds are commercially available. Among all manufacturing costs, drug application and vaccinations cover about 2% [11]. In poultry production, the use of hormones is prohibited and is not considered profitable [2].

Antimicrobials are widely used for the disease prevention and treatment, sustain the health in all poultry treated, induce growth, and enhance the quality of the meat for the purpose of reducing production costs. In European Union (EU), the usage of antimicrobials for the promotion of growth has been prohibited since 2006; whereas in United States, it is still currently been used for this purpose [12]. In addition to their specific effects, edible tissues of poultry might contain veterinary drug residues, which would cause hazardous health effects in human, such as direct toxicological/pharmacological effects, hypersensitivity, allergic reactions, change of gut microflora, and increased bacterial resistance to antibacterials [13]. Serious concerns are raised on the antibacterial resistance in zoonotic enteropathogens (*Salmonella* spp., *Campylobacter* spp.), commensal bacteria (*Escherichia coli, Enterococci*), and bacterial pathogens of animals (*Pasteurella, Actinobacillus* spp.) [4].

Nowadays, the prevention of coccidiosis has become the norm in modern farming for broilers and turkeys. Therefore, broiler producers almost constantly contain a coccidiostat in grower rations for reduced morbidity and mortality. Ionophores are the most extensive used drugs for the prevention of coccidiosis. Some drugs are not well absorbed from alimentary canal or not detrimental to require a withdrawal period, yet they can be used until slaughter legally. However, the majority of coccidiostats require withdrawal periods. Currently, vaccination is popular for the control of coccidiosis, which thereby eliminates the risk of residue transfer into poultry products [14].

Drug residues in poultry eggs are an issue of concern since only a few drugs are approved for laying hens, while a variety of drugs are approved for other production types. Residues could be accidental through mixing the feed in the same mill with the previously medicated feed or off-label treatment [12].

A large part of integrated poultry breeders monitor potential residues in meat or eggs routinely. This practice decreases possibility of drug residues in tissues, while the variation in cost among withdrawal feed and grower feed is significant. Some businesses do further monitoring in tissue before slaughter [9]. Fat and other tissues are analyzed for residues of contaminants such as pesticides, toxic elements, or persistent organic pollutants. Compared to fruit and vegetables, residues of pesticides such as organophosphorus, fungicides, herbicides, and carbamate compounds in poultry meat and products are negligible due to the elimination of these compounds to certain extent [12, 15]. Therefore, the risk of adverse health effects on human is assumed to be relatively low. In order to reduce adverse health risks arising from residues, Food and Agriculture Organization (FAO) and World Health Organization (WHO) expressed the mandate regarding the legal withdrawal periods for each drug [16, 17].

Fast and successful treatment of poultry infectious diseases is essential since it might lead important economical losses [18]. Therefore, selective, fast-acting, and potent drugs are selected [10]. So the drug selection at treatment of poultry infectious diseases, arrangement of treatment programs, determination of application ways, determination of individual or collective treatment doses and periods, and finally treatment of clinic efficiency of the drug are very different compared to the treatment options specific to the animal species [19].

Veterinary drugs are generally added to the feed as "feed additive" in the feed factories [20]. For the proper dosage, homogeneity of the active compound in the feed has great importance. Homogeneity of the medicated feed is standardized using specialized mixing equipments, yet these devices are expected to be cleaned each and after the preparation of the medicated feed. Documentation of procedures and relevant records are essential for each feed lot for legal requirements and external inspections. In order to prevent cross contamination, medicated feed should be stored in separate and not in the poultry house; where the bins or silos in the storage areas should also be cleaned properly [9].

3. Toxic elements

Entry of undesirable substances into the food chain is mainly due to environmental pollution. Fortunately, eggs are not a significant source of toxic elements, since only negligible amounts of these are able to penetrate the egg. Poultry meat may be contaminated with toxic elements such as arsenic, cadmium, or lead as a result of coming into contact with the materials on the farm or factory or while moving through marketing channels. These three toxic elements are known to induce more/widespread adverse health effects, which would be emphasized [21, 22].

3.1. Arsenic (As)

Organic arsenic compounds (roxarsone, arsanilic acid, nitarsone, and carbarsone) have been widely used in the poultry sector for long years as they prevent the diseases, accelerate growth, increase feed efficiency, and increase pigmentation of the meat. More than 90% of these organic arsenic compounds that are given to the chicken as feed additive are excreted with the feces as unchanged. The manures prepared from chicken feces including arsenic are applied to the croplands in order to increase efficiency of the soil; which would eventually lead environmental pollution [23]. Along with that, organic arsenic compounds available at the chicken manure were found to transform into dimethyl arsenic acid, monomethyl arsenic acid, and inorganic arsenic compounds; which are even more toxic than the ones available in the environment. Also, residuals of arsenic compounds were seen at body fat, liver, egg, and feather of the chicken fed with organic arsenic compounds [24]. Therewith, the use of organic arsenic compounds in 1998 was forbidden in EU countries. USA Food and Drug Administration (FDA) forbids the arsanilic acid, carbarsone, and roxarsone use in 2012 and nitarsone in 2015 [25].

3.2. Lead (Pb)

Due to pressure of the concerns in public health along with the regulations, a decrease in Pb emission along with the developments in the quality of chemical production in the recent years significantly decreased the Pb content in the environment. Even though this decrease, Pb, was still found in many food products such as giblets and offal at low concentrations [26]. Processing or production of foods in the fields contaminated with Pb was found to increase Pb level at the foods. Recent researches reveal that chronic Pb intoxication with low concentrations were found to cause pain, constipation, anemia, and an increase in hypertension and cardiovascular diseases in adults, while neuropathological disorders and even learning capacities are affected in children [27]. It was defined in Codex Alimantarius that Pb levels at a maximum of 0.1 mg/kg in the poultry meat and 0.5 mg/kg in edible offal were found to cause no adverse health effects [28].

3.3. Cadmium (Cd)

It is a metal that come to forefront as an environmental contaminant resulted from both natural and industrial and agricultural sources. The individuals who do not smoke are exposed to Cd through foodstuffs [29]. Absorption of Cd through digestive canal is very low in the humans (3–5%); while they are able to accumulate in liver and kidney at significant amounts in human. Its biological half-life is very long (10–30 years). Cd mainly leads to damage of kidney functions by damaging proximal tubular of the kidneys [30]. Also it leads to bone damage both directly and indirectly. The tubular damage is seen as a result of exposure to Cd at low dose for a long period or at high dose for a short period. As a result, glomerular filtration speed is decreased and finally renal impairment is seen [31]. International Agency on Cancer Research (IARC) classifies Cd as human carcinogenic (Group 1) [32]; while European Union Food Safety Authority (EFSA) also underlined this fact stating, and exposure of Cd would lead increased risk for lung, endometrium, urinary bladder, and breast cancer [33].

In order to avoid the bioaccumulation and for the sake of public health, authorities are obliged to conduct surveillance and monitoring analysis for metal contamination in poultry. It has been understood from the studies and assessments concerning to the metal pollution at the poultry meats that metal pollution that may be available at the poultry meats do not lead to a significant public health concern but it is required to be avoided from the offal of elder animals as a precaution [21, 34].

4. Radioactive substances

Ionizing radiation is a natural source of energy as people are exposed through soil, water, and vegetation. The radio waves that allow for radio and television communication, X-rays used in the medicine and industry, and solar rays are the radiation types that we are accustomed in our daily life. Humans and animals are exposed to the radioactive substances through various sources mainly through air, water, and alimentary by food feed. Nuclear trials, nuclear power plants, effluents, and residuals of the nuclear researches, nuclear accidents, mine pits including radioactive substances, facilities producing radioisotope, radioisotopes used in the scientific researches, electron microscopes, and radioactive rays used in the medicine and industry are the main sources of contamination for environment as well as food [35].

There are so many disasters in the history regarding nuclear accidents, nuclear trials, and nuclear leakages. These substances may be carried to far away from the places that they are located through air and water. As they are resistant physically, chemically, and biologically, most of them enter into food chain, leading long-term permanent undesirable effects [36]. These radioactive compounds taken through inhalation, dermal, and alimentary routes are found to accumulate at the tissues and organs, which would cause to damages to the surrounding cells, tissues, and organs by their particle composition or by emitting rays (internal ray) [37]. Intoxication with radiation with these internal radiation emitting substances may progress as acute or chronic. Chronic effects of the radiation, available at the water and foodstuffs at low amounts, are more important in the living creature in terms of food toxicology [38].

The nuclear trials, nuclear accidents, nuclear leakages, and nuclear contaminations as a result of use of nuclear weapons are the most significant sources of radioactive substance contamination in the foodstuffs. With the help of various air movements, the substances emitted to the places close to or far away from these sources contaminate plants, agricultural products, fruits, or waters; they enter into the bodies of animals that crop or the humans live in these places as well as various food sources. Also, they are transferred into the consumers through the foodstuffs obtained from these animals [39].

The researches concerning to the radioactive substances at the animals of food origin started in 1950–1960s with the tests of atomic bombs. The Chernobyl accident carried out a better warning duty in terms of understanding the factors that lead to transfer of radioactive substances to the animals and allowed for development of more appropriate precautions. International Atomic Energy Agency published a booklet including the precautions to be taken concerning to the radioactive substances available at the foods of animal origin [40]. Iodine-131 and cesium-137 were found as radioactive residues in food following Chernobyl, the biggest nuclear accident in Pripyat, Ukrainian SSR. Even after 30 years, current researches still emphasis the remedies of the residues of Chernobyl along with the recent nuclear disaster Fukushima, Japan, which is the second biggest nuclear accident.

4.1. Iodine-131 (^{131}I).

Iodine-131 is available at the herbs in the meadows and forages and transferred to the animals through the feed that were cropped in these areas. It is fully absorbed from the digestive canal and concentrates on the thyroid gland. It is transferred to animal products at differing rates, such as for milk at 6% [41] and for egg 15% [42]. It was reported that ^{131}I is still found in chicken eggs after single exposure for 20 days; while in case of a longer exposure such as exposure longer than 1 week, residues may still be found up until 30 days. Therefore, the withdrawal period for eggs exposed to ^{131}I contaminated chicken feed was defined as 30 days [42].

4.2. Cesium-137 (^{137}Cs)

This substance is especially resulted from nuclear weapons, nuclear reactor leakages and residuals. Absorption of ^{137}Cs from the digestive canal varies between <10 and 100% in accordance with its chemical form, which is mostly distributed in the soft tissues of the body. As it is firmly bound to the soil, it may not enter into the food chain easily; but because its disengaged part at the soil is too much at the locations where it is common, it may be transferred to the plants and therefore animals and their products [43]. Its withdrawal speed from the muscle tissue is much more at the small animals compared to large animals. For example; ^{137}Cs is removed from the muscle tissue in 1–2 days at the chicken, while this period may extend up to 60 days for calves [44]. Even though the distribution of ^{137}Cs at the livestock (pig, sheep, layer hen, and broilers) varies after given orally, it was notified that the highest concentration was found at muscle and kidneys regardless the animal type. The lowest ^{137}Cs concentration was determined at the blood. The absorption through digestive canal after given orally were found to be fast in broilers; where the highest concentrations were found at chest, leg, intestine, and liver for the first day. Meanwhile, the highest accumulation was seen at the muscle tissue at the layer hens [45].

5. Persistent organic pollutants (POPs)

POPs is the general definition for the natural or synthetic organic compounds which are known to bioaccumulate in the environment without being degraded due their resistance to chemical, biological, and optical breakdown and have detrimental effects to human, animal, plant, and environment. The traces of these substances are not restricted to the areas where the production or application exists, but also with their stability which would lead long-range transport they are even found in uninhabitant areas where POPs have never been used. This issue raised a global concern and initiated global measures for the prevention of the pollution [46].

POPs have low water solubility and high lipid solubility (oil, fat) with low steam pressure, which makes their half-life in the environment long. Due to these properties, they accumulate on the fat tissues of the living creatures including human. Therefore, both acute and chronic toxic effects occur in human, wild animals, and other organisms, which are exposed to POPs for many generations creating bioaccumulation [47]. Human are also exposed to POPs by food and in utero during gestation or infant by lactation periods. Several studies proved that POPs induce endocrine disruption and effect immune, nervous and reproduction systems, even causing cancer [48].

As these compounds are used and manufactured in various sectors of the industry and agriculture; they might be formed as a consequence of by-products or even at burning procedures [49, 50]. Among these formed unintentional by-products, dioxins are the most known. The sources that lead to dioxin and other POP residuals in poultry are as follows [51]:

– Volatilization of these compounds to the air from the undefined burning procedures in industry and accumulation in the soil,

– Use of some clay minerals such as kaolin or clay ball,

– The waste disposal regions where materials including Polychlorobiphenyls (PCBs) are buried,

– The red slag (Kieselrot) formation during the copper production, polychlorobenzodioxins/furans (PCDD/Fs) used in surfacing of the road surfaces and playgrounds at significant amounts,

– Fall of the isolation material at the roof of poultry house that is decomposed in time and its intake by the animals,

– The grouting in the buildings.

Chicken are exposed to POPs through various ways. The most important exposure route is its intake through feed. POP residuals at the feeds come from feed raw materials including fat. Along with alimental exposure, chicken are also exposed to POPs through dermal and inhalation routes Araclor 1254 (PCB mixture) given by feed at 0.5–1 ppm concentration was found to transfer the egg at 0.2–0.45 ppm levels. Meanwhile, administration of the same compound to Leghorn chicks for 8–32 weeks at 0.1–10 ppm, no toxic effect was observed including egg efficiency and hatching ratio. If Araclor 1254 and other PCB mixtures (Araclor 1248, 1242 and 1232) exceeds over 20 ppm, a decrease of the egg efficiency and hatching ratios along with teratogenic effects were observed [52, 53]. The finding in this study were found in accordance to another study where Araclor 1248 at 0.5–1 ppm given through feed, did not affect egg efficiency and hatching; while 10–20 ppm significantly decreased the egg efficiency in 8 weeks [54].

As a conclusion, chicken bred in extensive poultry production systems are more exposed to infectious agents and chemical contaminants compared to intensive systems. This susceptibility creates an increased risk of diseases and brings the fact of the increased use of veterinary drugs, which would eventually cause residues in the edible tissues and other products in poultry [51].

6. Polycyclic aromatic hydrocarbons (PAHs)

PAHs are the chemical compounds, which includes two or more aromatic rings and consist of carbon and hydrogen atoms which compromise pyrolysis or partial burning of the organic substances due to the industrial proceedings and human effects. The processing processes of the foods (smoking and drying) and cooking at high temperatures (grilling, frying, and roasting) creates the main source of PAHs [55]. Pyrolysis occurs due to the drip of the oil to the flame while cooking the foods at above 200°C and PAHs contaminate the foods along with the smoke that will occur [56]. PAH levels occurred as a result of applying different cooking methods to the fowl for 0.5–1.5 h were compared, where the highest PAH content was found to be formed when the meat was cooked at the coal flame (320 μg/kg). This was followed by cooking with the skin at the coal flame (300 μg/kg), smoking (210 μg/kg), frying (130 μg/kg), steaming (8.6 μg/kg), and liquid smoking (0.3 μg/kg) [57]. In the same way, PAH was found to be produced at the highest, when the meat was directly subjected to the wood fire and at the lowest level when it was cooked at the cinder. With the disintegration occurred as a result of drip of melted fats on the heat source during cooking the meats, formation of PAHs was found to increase and volatilized to the atmosphere, accumulate back again on the meat. PAHs are mainly lipophilic and mostly stored at fat tissue, where an increase of fat content of the meat, would directly lead to increase of PAH amount. PAH contents were also found to be significantly increased as a result of cooking hamburger, beef, fish, and chicken meats at the high temperature barbecue. The coal and coal dusts at the barbecue penetrates on the cooked foods and make carcinogenic effect and mainly cause to alimentary canal and large intestine cancers [58].

Smoking is one of the oldest technologies that have been used for maintaining meats and meat products and defined as penetration procedure of the steams formed with degradation of the wood with the heat to the meat products. In general, it is widely used at the processing the fishes. PAHs may occur during insufficient burning of the wood as an undesired conclusion of the smoking. There are approximately 660 different compounds within PAH group, in which some of these have carcinogenic properties [59]. Benzo[a]pyrene (BaP), which is the most known carcinogenic PAH compound, has been used as lead coating material up to date. In EU legislation, maximum residual limit of BaP is 5 μg/kg at fume (smoked) meat and meat products. Also European Commission suggests that the member countries should research not only BaP amount but also 15 other PAHs which have possible carcinogenic effects at cured meats. Dibenzo[a,i]pyrene (DiP) received attention, due to the recent toxicological studies showing its potency in carcinogenicity much greater than BaP. EFSA also suggests the analysis of benzo[c]fluorene (BcL) since The Joint FAO/WHO Expert Committee on Food Additives (JECFA) considered at particular interest; while specific scientific studies are still missing [60].

7. Phthalates

Phthalates are the chemical substances that are used for making the plastic materials softer, more elastic, lighter, and more resistant. These kinds of plastics are also used for keeping the foodstuffs fresh, maintaining curative effects of continuous release pharmaceutical substances, and preventing burning or spread of the fire of electronics and other household and cosmetic products [61]. These compounds are produced at very high amounts since 1930s and their production amount globally was found as 4.9 million tons in 2010. Therefore, residues of phthalates such as dimethyl phthalate (DMP), diethyl phthalate (DEP), di-n-propyl phthalate (DPP), diisobutyl phthalate (DiBP), di-n-butyl phthalate (DBP), butyl benzyl phthalate (BBzP), dicyclohexyl phthalate (DCHP), diisoheptyl phthalate (DHP), di(2-ethylhexyl) phthalate (DEHP), diisodecyl phthalate (DiDP), and diisononyl phthalate (DiNP) are inevitable and widely seen at the foods and food packages [62].

Phthalates are defined among the endocrine disrupting chemicals due to their effects on the reproduction system. They found to cause sperm damage, early puberty at the females, reproduction canal abnormalities, early pregnancy termination, or hepatic tumors in rodents [63]. Widely existence of plastic materials in the nature, interest to personal care products, and various food packaging shows that humans and animals are indispensably exposed to these chemicals. Phthalates are taken through mouth, breathing, injection, or dermal contact [64, 65]. According to USA Centers for Disease Control and Prevention (CDC) reports, almost all of the humans screened were found to have phthalate metabolites at detectable concentrations in their urine. The levels of phthalates metabolytes in urine were linked to the possible daily routines of the subjects, which shows that the phthalate metabolites of adult women using soaps, body washing liquids, shampoos, cosmetic, and similar products much higher compared to the men [66]. Along with the environmental cosmetic exposure, food products such as milk, butter, and meats are very common sources for some phthalates such as DEHP, DBP and DiBP due to that they have lipophilic character [67].

Based on these reports, continuous screening of these phthalates in foods and food products has great importance. The contamination could be from the environment as well as the plastic materials used in the production stages. Even though their usage in food packaging is decreased, many products include phthalates for softening purposes as well as various processing stages [68].

Phthalates may penetrate the foods during production, packing and preparation. Due to wide usage and availability in various products, including the laboratory materials, analysis of the food is difficult. In a study, nine phthalate esters in 72 different foods commonly consumed were screened and DEHP was found at 74% of all these products including the infant foods [69].

On the contrary to other chemical contaminants, information regarding phthalates as residue in food is very limited. Since their usage with the improved technology, lead mixed contamination at very low concentrations; previous studies do not reflect the current status. The tolerable daily intake for some phthalates was stated as 0.01, 0.5, 0.05, 0.15, and 0.15 mg/kg/day for DBP, BBzP, DEHP, DiNP, and DiDP, respectively by EFSA [68, 70].

The habitats and feeding ways of the chicken that are free-range extensive and intensive or semi-poultry houses may make difference on their exposure to the environmental pollutants such as phthalates. Information regarding the indoor-outdoor breeding on phthalate levels is missing, except one study showing only the levels of phthalates in chicken eggs in retail. This research showed that DEHP may even be found at the shell of membrane of the egg [71] and DBP and DEHP might be present at even low concentrations [72]. The same study revealed that DBP and DEHP were found only in egg white as 0–0.15 and 0.05–0.4 ppm, while no contamination was reported for egg yolk. Meanwhile another experimental study showed that DEHP may be present both at white yolk and white after it is given to the chicken orally at single and repeated dose applications (3 days). The animals are also given DEHP for 45 days at a concentration of 1/100 g feed and the results of the transfer to the egg yolk was found to be even more in these fed animals [73]

8. Other contaminants

There are many other compounds that may cause adverse health effects through poultry and poultry products. WHO and American Dietetics Association identified the potential hazards of trans-fatty acids on human health, which are formed from further-processed, deep-fried products from fast-food retails. Fatty acids in oils are generally in the cis-form, which will be transformed to the less healthy trans-form during hydrogenation and over-heating during deep frying. Compared to pesticide residues, trans-fatty acids also lead deleterious effects; which were found to be directly related to atherosclerosis, cardio-vascular diseases, cancer, ulceration, and oxidative degeneration [74].

9. Control of the contaminants at poultry meat and products

The control strategies for the pharmacological effective substances and environmental pollutants in poultry meat and products includes complex and evolving measures for defining/determination of these compounds and analysis of exposure/prevention routes.

Recent advances in analytic chemistry lead to the understanding of chemical contamination frequency/routes/prevention mechanisms in the food field more intensively. While some of these threat the health even at low concentrations at no threshold, some of them may be have legal breaching characteristic [75]. These require various risk method approaches. For appropriate control of the production procedures, biosafety, full traceability, good hygiene, and other sanitary applications are compulsory requirements [7].

As well as that the contaminants at the foodstuffs may be determined with special chromatographic methods, they may be also determined with very sensitive and cheap methods as described for antibiotics detection. The residues are found at blood, urine, sometimes gall, and mostly edible tissues. The blood, sometimes gall, muscles, liver, and egg are the target matrices that may be analyzed [76]. The amount to be analyzed is generally 1 ppm and sometimes below than 1 ppb. These analyses are based on degradation of the residuals following extraction of the residuals from poultry tissues with water or organic solvents or chromatography procedures. The disintegration is done with, immiscible liquids such as water and petroleum or two environments such as liquid and solid [7].

The analytic methods that are used for monitoring antibiotic residuals are generally classified into two groups such as "verification methods" and "screening methods." The verification methods are generally based on liquid chromatography combined with mass spectrophotometer (LC/MS) for determining the concentration of the analyte. Sometimes, the verification procedure may be carried out with the methods based on LC with ultraviolet (UV) detector or capillary electrophoresis (CE). Along with that, all of these require time-consuming, expensive and complicate laboratory equipments and trained personnel [75]. Also they are subjected to solid phase extraction (SPE) or very demanding sample preparation procedures based on multistep cleaning for extraction of the analyte. The screening methods may determine the analyte. But generally, semi-quantitative results are obtained. Ideal properties of a screen method are as follows: they give very less wrong positive results, they are efficient, their use is simple, their analyses are short, and they are selective and low priced [77].

In the current literature, the most widely used techniques in the analysis of the antibiotics are LC/MS (at the rate of 38%) and LC/UV (at the rate of 18), which are the verification methods, and ELISA method (at the rate of 18), which is a screen method. Along with that, an increase of other screening methods (at the rate of 12%) and use of biosensors (at the rate of 8%) are available. As there are so many and various types of drug residuals in the animal products, many prior screening analyses are required to be completed for an effective verification; yet therefore, the use of screen tests is compulsory. Since it is cheap and does not require complicated devices and advanced trained scientists, screening methods are applied by poultry farmers extensively in the field [78].

Microbiological methods are qualitative or semi-qualitative methods that are based on the reaction between sensitive bacteria and antibiotic at the sample. The advantages of these methods are their simplicity, reliability and their low price. Various tests are commercially available. Meanwhile microbiological methods are less sensitive, LC/MS methods may only be applied to the component to be selected as the target. In this way, any other possible antibacterial substance may not be seen. Along with that, the most important disadvantages of the microbiological methods are that they have selectivity and sometimes they need for a long incubation period. In the microbiological methods that are used for determining the residuals of the antibiotics at the solid foodstuffs, the procedures based on simple solid-liquid extraction (SLE) are the most preferred sample preparation procedure. Pollution at the processing the samples and solid phase extraction (SPE) at the degreasing or liquid-liquid extraction with hexane may be used [79].

Another most preferred screen test is the use of ELISA systems widely available for various kinds of contaminants such as hormones and drugs. This method has a very high specificity and due to use of antibodies that is specifically developed for the target molecule, it is assumed as very sensitive. Due to the specificity and reliability, analysis of different residuals at a very short time instantly is available along with relatively easy sample preparation procedures; ELISA methods are assumed as a "must" in poultry field. A typical ELISA kit is formed from 96 wells coated with the antibodies developed for target compounds, where sample and standards are added at specific amounts and other steps of reaction and washing are followed. The amount is assigned by a simple microplate reader [80].

While some liquid samples (such as urine and plasma) are directly analyzed by diluting with a buffered solution, solid samples such as meat and egg are extracted using liquid-liquid extraction by organic solvents or solid-liquid extraction using special colons followed by cleaning procedures. In some cases, chemical procedures are needed for disintegration of target molecule such like in the nitrofurans [78].

The screening tests are used for monitoring the slaughter houses, control of the import, control of the feeds, or market controls. In cases of acceptance or rejection of a feed lot, the decision may be given in accordance with ELISA results; while in the cases including legal sanctions, the positive results obtained from ELISA are required to be verified with the methods at which expensive and sophisticated laboratory equipments are used and that require difficult sample preparation stages. These analyses are carried out by the experts and strict quality control parameters are applied with the aim of achieving reliability of the results [81].

10. Conclusions

Today, the use of emerging technologies at all stages of food production, the growth of the fast-food industry, and environmental pollution lead to increase the risk of contamination of food. Home cooking and food preparation habits have been gradually decreased; while nowadays readily available convenience foods or chain restaurants are preferred. In order to serve as much people possible, companies or other food chain industries generally retail raw food as bulk from market or poultry producers with a possibility of veterinary drug residues or other environmental contaminants at very high concentrations. Even, machine washing and cleaning in restaurants may present less efficacy leading to more possible surface contaminants. Nevertheless, food manufacturers are needed to perceive HACCP training for the emerging contaminants and consider food safety regulations and follow Good Manufacturing Practices. There have to be rigorous improved quality controls by exactly implementing the HACCP program at every step of food production and processing for these contaminants.

Author details

Ayhan Filazi[1*], Begum Yurdakok-Dikmen[1], Ozgur Kuzukiran[2] and Ufuk Tansel Sireli[3]

*Address all correspondence to: filazi@veterinary.ankara.edu.tr

1 Department of Pharmacology and Toxicology, Faculty of Veterinary Medicine, Ankara University, Ankara, Turkey

2 Veterinary Control Central Research Institute, Ankara, Turkey

3 Department of Food Hygiene and Control, Faculty of Veterinary Medicine, Ankara University, Ankara, Turkey

References

[1] Weber R, Watson A, Forter M, Oliaei F. Persistent organic pollutants and landfills – a review of past experiences and future challenges. Waste Manag Res. 2011;29:107–121. doi:10.1177/0734242X10390730

[2] Sireli UT, Filazi A, Onaran B, Artik N, Ulker H. Residual concerns in meat. Turkiye Klinikleri J Food Hyg Technol-Special Topics. 2015;1:7–16.

[3] Reyes-Herrera I, Donoghue DJ. Chemical contamination of poultry meat and eggs. In: Schrenk D, editor. Chemical contaminants and residues in food. A volume in Woodhead Publishing Series in Food Science, Technology and Nutrition, Cambridge, UK. 2012; p. 469–497.

[4] Di Stefano V, Avellone G. Food contaminants. J Food Studies. 2014;3:88–102. doi: 10.5296/jfs.v3i1.6192

[5] Codex Stand 193. Codex General Standard for Contaminants and Toxins in Food and Feed [Internet]. 1995. Available from: http://www.fao.org/fileadmin/user_upload/agns/pdf/CXS_193e.pdf [Accessed: 2016 April 30].

[6] Andree S, Jira W, Schwind KH, Wagner H, Schwägele F. Chemical safety of meat and meat products. Meat Sci. 2010;86:38–48. doi:10.1016/j.meatsci.2010.04.020

[7] Botsoglou NA, Fletouris DJ. Drug Residues in Food. Marcel Dekker, Inc., New York, NY. 2001, p. 269–298.

[8] Marshall BM, Levy SB. Food animals and antimicrobials: impacts on human health. Clin Microbiol Rev. 2011;24:718–733. doi:10.1128/CMR.00002-11

[9] Filazi A. Antibiotic residues in food of animal origin and evaluating of their risks. Turkiye Klinikleri J Vet Sci. 2012;3:1–7.

[10] Filazi A, Yurdakok-Dikmen B, Kuzukiran O. Antibiotic resistance in poultry. Turkiye Klinikleri J Vet Sci Pharmacol Toxicol-Special Topics. 2015;1:42–51.

[11] National Research Council. The use of drugs in food animals: benefits and risks. Committee on Drug Use in Food Animals, Panel on Animal Health, Food Safety, and Public Health. National Academy Press, Washington, DC. 1999, p. 27–68.

[12] Goetting V, Lee KA, Tell LA. Pharmacokinetics of veterinary drugs in laying hens and residues in eggs: a review of the literature. J Vet Pharmacol Therap. 2011;34:521–556. doi:10.1111/j.1365-2885.2011.01287.x

[13] Beyene T. Veterinary drug residues in food-animal products: its risk factors and potential effects on public health. J Veterinar Sci Technol. 2016;7:285. doi: 10.4172/2157-7579.1000285

[14] Arabkhazaeli F, Nabian S, Modirsanei M, Madani SA. The efficacy of a poultry commercial anticoccidial vaccine in experimental challenge with Eimeria field isolates. IJVM. 2014;8:249–253.

[15] Aulakh RS, Gill JPS, Bedi JS, Sharma JK, Joia BS, Ockerman HW. Organochlorine pesticide residues in poultry feed, chicken muscle and eggs at a poultry farm in Punjab, India. J Sci Food Agr. 2006;86:741–744. doi:10.1002/jsfa.2407

[16] Filazi A, Sireli UT, Cadirci O. Residues of gentamicin in eggs following medication of laying hens. Br Poult Sci. 2005;46:580–583. doi:10.1080/00071660500273243

[17] Filazi A, Sireli UT, Yurdakok B, Aydin FG, Kucukosmanoglu AG. Depletion of florfenicol and florfenicol amine residues in chicken eggs. Br Poult Sci. 2014;55:460–465. doi:10.1080/00071668.2014.935701

[18] Filazi A, Sireli UT, Pehlivanlar-Onen S, Cadirci O, Aksoy A. Comparative pharmacokinetics of gentamicin in laying hens. Kafkas Univ Vet Fak Derg. 2013;19: 495–498. doi: 10.9775/kvfd.2012.8138

[19] Landoni MF, Albarellos G. The use of antimicrobial agents in broiler chickens. Vet J. 2015;205:21–27. doi:10.1016/j.tvjl.2015.04.016

[20] Armut M, Filazi A. Evaluation of the effects produced by the addition of growth-promoting products to broiler feed. Turk J Vet Anim Sci. 2012;36:330–337. doi:10.3906/vet-1010-2

[21] Kurnaz E, Filazi A. Determination of metal levels in the muscle tissue and livers of chickens. Fresen Environ Bull. 2011;20:2896–2901.

[22] Sanap MJ, Jain N. Cadmium profile in visceral organs of experimentally induced toxicity in Kadaknath chicken. Environ Ecol. 2015;33:807–809.

[23] Mangalgiri KP, Adak A, Blaney L. Organoarsenicals in poultry litter: detection, fate, and toxicity. Environ Int. 2015;75:68–80. doi:10.1016/j.envint.2014.10.022

[24] Fisher DJ, Yonkos LT, Staver KW. Environmental concerns of roxarsone in broiler poultry feed and litter in Maryland, USA. Environ Sci Technol. 2015;49:1999–2012. doi:10.1021/es504520w

[25] Baynes RE, Dedonder K, Kissell L, Mzyk D, Marmulak T, Smith G, Tell L, Gehring R, Davis J, Riviere JE. Health concerns and management of select veterinary drug residues. Food Chem Toxicol. 2016;88:112–122. doi:10.1016/j.fct.2015.12.020

[26] Ersoy IE, Uzatıcı A, Bilgücü E. Possible heavy metal residues in poultry and their products that are bred around cement industry. JABB. 2015;3:63–68. doi:10.14269/2318-1265/jabb.v3n2p63-68

[27] Petit D, Véron A, Flament P, Deboudt K, Poirier A. Review of pollutant lead decline in urban air and human blood: a case study from northwestern Europe. C R Geoscience. 2015;347:247–256. doi:10.1016/j.crte.2015.02.004

[28] Ismail SA, Abolghait SK. Estimation of Lead and Cadmium residual levels in chicken giblets at retail markets in Ismailia city, Egypt. Int J Vet Sci Med. 2013;1:109–112. doi:10.1016/j.ijvsm.2013.10.003

[29] Akerstrom M, Barregard L, Lundh T, Sallsten G. Variability of urinary cadmium excretion in spot urine samples, first morning voids, and 24 h urine in a healthy non-smoking population: implications for study design. J Expo Sci Environ Epidemiol. 2014;24:171–179. doi:10.1038/jes.2013.58

[30] Bernhoft RA. Cadmium toxicity and treatment. Sci World J. 2013; Article ID 394652, 7 pages.

[31] Nair AR, Smeets K, Keunen E, Lee WK, Thévenod F, Van Kerkhove E, Cuypers A. Renal cells exposed to cadmium *in vitro* and *in vivo*: normalizing gene expression data. J Appl Toxicol. 2015;35:478–484. doi:10.1002/jat.3047

[32] IARC. Cadmium and Cadmium Compounds. IARC Monographs – 100C [Internet]. 2012. Available from: http://monographs.iarc.fr/ENG/Monographs/vol100C/mono100C-8.pdf [Accessed: 2016 April 30].

[33] EFSA. Scientific opinion: cadmium in food. Scientific opinion of the panel on contaminants in the food chain. EFSA J. 2009;980:1–139.

[34] Khalafalla FA, Abdel-Atty NS, Abd-El-Wahab MA, Ali OI, Abo-Elsoud RB. Assessment of heavy metal residues in retail meat and offals. J Am Sci. 2015;11:50–54.

[35] Kjaer A, Knigge U. Use of radioactive substances in diagnosis and treatment of neuroendocrine tumors. Scand J Gastroenterol. 2015;50:740–747. doi:10.3109/00365521.2015.1033454

[36] Borron SW. Introduction: Hazardous materials and radiologic/nuclear incidents: lessons learned? Emerg Med Clin North Am. 2015;33:1–11. doi:10.1016/j.emc.2014.09.003

[37] Weisdorf D, Chao N, Waselenko JK, Dainiak N, Armitage JO, McNiece I, Confer D. Acute radiation injury: contingency planning for triage, supportive care, and transplantation. Biol Blood Marrow Transplant. 2006;12:672–682. doi:10.1016/j.bbmt.2006.02.006

[38] Yamaguchi K. Investigations on radioactive substances released from the Fukushima Daiichi nuclear power plant. Fukushima J Med Sci. 2011;57:75–80. doi:10.5387/fms.57.75

[39] Merz S, Steinhauser G, Hamada N. Anthropogenic radionuclides in Japanese food: environmental and legal implications. Environ Sci Technol. 2013;47:1248–1256. doi:10.1021/es3037498

[40] Steinhauser G, Brandl A, Johnson TE. Comparison of the Chernobyl and Fukushima nuclear accidents: a review of the environmental impacts. Sci Total Environ. 2014;470–471:800–817. doi:10.1016/j.scitotenv.2013.10.029

[41] Treinen RM. An analysis of the intake of iodine-131 by a dairy herd post-Fukushima and the subsequent excretion in milk. J Environ Radioact. 2015;149:135–143. doi:10.1016/j.jenvrad.2015.07.017

[42] Unak T, Yildirim Y, Avcibasi U, Cetinkaya B, Unak G. Transfer of orally administered iodine-131 into chicken eggs. Appl Radiat Isot. 2003;58:299–307. doi:10.1016/S0969-8043(02)00350-0

[43] Murakami M, Ohte N, Suzuki T, Ishii N, Igarashi Y, Tanoi K. Biological proliferation of cesium-137 through the detrital food chain in a forest ecosystem in Japan. Sci Rep. 2014;4:3599. doi:10.1038/srep03599

[44] Mitrovic BM, Vitorovic G, Vicentijevic M, Vitorovic D, Pantelic G, Lazarevic-Macanovic M. Comparative study of ^{137}Cs distribution in broilers and pheasants and possibilities for protection. Radiat Environ Biophys. 2012;51:79–84. doi:10.1007/s00411-011-0391-8

[45] Beresford NA, Howardy BJ. An overview of the transfer of radionuclides to farm animals and potential countermeasures of relevance to Fukushima releases. Integr Environ Assess Manag. 2011;7:382–384. doi:10.1002/ieam.235

[46] Kuzukiran O, Yurdakok-Dikmen B, Filazi A, Sevin S, Aydin FG, Tutun H. Determination of polychlorinated biphenyls in marine sediments by ultrasound-assisted isolation and dispersive liquid-liquid microextraction and gas chromatography-mass spectrometry. Anal Lett. doi:10.1080/00032719.2016.1151890

[47] Yurdakok B, Tekin K, Daskin A, Filazi A. Effects of polychlorinated biphenyls 28, 30 and 118 on bovine spermatozoa *in vitro*. Reprod Domest Anim. 2015;50:41–47. doi:10.1111/rda.12447

[48] Kuzukiran O, Filazi A. Determination of selected polychlorinated biphenyl residues in meat products by QuEChERS method coupled with gas chromatography-mass spectrometry. Food Anal Method. 2016;9:1867–1875. doi:10.1007/s12161-015-0367-4

[49] Filazi A, Yurdakok-Dikmen B, Kuzukıran O. Poisoning cases originated from environmental pollutants. Turkiye Klinikleri J Vet Sci Pharmacol Toxicol-Special Topics. 2015;1:45–52.

[50] Yurdakok-Dikmen B, Kuzukiran O, Filazi A, Kara E. Measurement of selected polychlorinated biphenyls (PCBs) in water via ultrasound assisted emulsification–microextraction (USAEME) using low-density organic solvents. J Water Health. 2016;14:214–222. doi:10.2166/wh.2015.177

[51] Schoeters G, Hoogenboom R. Contamination of free-range chicken eggs with dioxins and dioxin-like polychlorinated biphenyls. Mol Nutr Food Res. 2006;50:908–914. doi:10.1002/mnfr.200500201

[52] Teske RH, Armbrecht BH, Condon RJ, Paulin HJ. Residues of polychlorinated biphenyl in products from poultry fed Aroclor 1254. J Agric Food Chem. 1974;22:900–904. doi:10.1021/jf60195a016

[53] Lillie RJ, Cecil HC, Bitman J, Fries GF. Differences in response of caged White Leghorn layers to various polychlorinated biphenyls (PCBs) in the diet. Poult Sci. 1974;53:726–732. doi:10.3382/ps.0530726

[54] Platonow NS, Reinhart BS. The effects of Polychlorinated Biphenyls (Aroclor 1254) on chicken egg production, fertility and hatchability. Can J Comp Med. 1973;37:341–346.

[55] Rozentāle I, Stumpe-Vīksna I, Začs D, Siksna I, Melngaile A, Bartkevičs V. Assessment of dietary exposure to polycyclic aromatic hydrocarbons from smoked meat products produced in Latvia. Food Control. 2015;54:16–22. doi:10.1016/j.foodcont.2015.01.017

[56] Simko P. Determination of polycyclic aromatic hydrocarbons in smoked meat products and smoke flavouring food additives. J Chromatogr B Analyt Technol Biomed Life Sci. 2002;770:3–18. doi:10.1016/S0378-4347(01)00438-8

[57] Farhadian A, Jinap S, Abas F, Sakar ZI. Determination of polycyclic aromatic hydrocarbons in grilled meat. Food Control. 2010;21:606–610. doi:10.1016/j.foodcont.2009.09.002

[58] Lee JG, Kima SY, Moona JS, Kima SH, Kang DH, Yoon HJ. Effects of grilling procedures on levels of polycyclic aromatic hydrocarbons in grilled meats. Food Chem. 2016;199:632–638. doi:10.1016/j.foodchem.2015.12.017

[59] Hitzel A, Pöhlmann M, Schwägele F, Speer K, Jira W. Polycyclic aromatic hydrocarbons (PAH) and phenolic substances in meat products smoked with different types of wood and smoking spices. Food Chem. 2013;139:955–962. doi:10.1016/j.foodchem.2013.02.011

[60] EFSA. Polycyclic aromatic hydrocarbons in food. Scientific opinion of the panel on contaminants in the food chain. EFSA J. 2008;724:1–114.

[61] Birnbaum LS, Schug TT. Phthalates in our food. Endocr Disruptors. 2013;1:1–5. doi: 10.4161/endo.25078

[62] Guo Y, Kannan K. Challenges encountered in the analysis of phthalate esters in foodstuffs and other biological matrices. Anal Bioanal Chem. 2012;404:2539–2554. doi: 10.1007/s00216-012-5999-2

[63] Yurdakok-Dikmen B, Alpay M, Kismali G, Filazi A, Kuzukiran O, Sireli UT. *In vitro* effects of phthalate mixtures on colorectal adenocarcinoma cell lines. J Environ Pathol Toxicol Oncol. 2015;34:115–123. doi:10.1615/JEnvironPatholToxicolOncol.2015013256

[64] Durmaz E, Ozmert EN, Erkekoglu P, Giray B, Derman O, Hincal F, Yurdakok K. Plasma phthalate levels in pubertal gynecomastia. Pediatrics. 2010;125:122–129. doi:10.1542/peds.2009-0724

[65] Kavlock R, Barr D, Boekelheide K, Breslin W, Breysse P, Chapin R, Gaido K, Hodgson E, Marcus M, Shea K, Williams P. NTP-CERHR expert panel update on the reproductive and developmental toxicity of di(2-ethylhexyl) phthalate. Reprod Toxicol. 2006;22:291–399. doi:10.1016/j.reprotox.2006.04.007

[66] CDC. Fourth Report on Human Exposure to Environmental Chemicals, Department of Health and Human Services Centers for Disease Control and Prevention [Internet]. 2009. Available from: http://www.cdc.gov/exposurereport/pdf/FourthReport.pdf [Accessed: 2016 April 30].

[67] Wormuth M, Scheringer M, Vollenweider M, Hungerbühler K. What are the sources of exposure to eight frequently used phthalic acid esters in Europeans? Risk Anal. 2006;26:803–824. doi:10.1111/j.1539-6924.2006.00770.x

[68] Cao XL. Phthalate esters in foods: Sources, occurrence, and analytical methods. Compr Rev Food Sci Food Saf. 2010;9:21–43. doi:10.1111/j.1541-4337.2009.00093.x

[69] Schecter A, Lorber M, Guo Y, Wu Q, Yun SH, Kannan K, Hommel M, Imran N, Hynan LS, Cheng D, Colacino JA, Birnbaum LS. Phthalate concentrations and dietary exposure from food purchased in New York State. Environ Health Perspect. 2013;121:473–494. doi:10.1289/ehp.1206367

[70] EFSA. Statement of the Scientific Panel on Food Additives, Flavourings, Processing Aids and Materials in Contact with Food on a request from the Commission on the

possibility of allocating a group-TDI for Butylbenzylphthalate (BBP), di-Butylphthalate (DBP), Bis(2-ethylhexyl) phthalate (DEHP), di- Isononylphthalate (DINP) and di-Isodecylphthalate (DIDP) [Internet]. 2005. Available from: http://www.efsa.europa.eu/sites/default/files/scientific_output/files/main_documents/747.pdf [Accessed: 2016 April 30].

[71] Suyama K, Nakamura H, Ishida M, Adachi S. Lipids in the exterior structures of the hen egg. J Agric Food Chem. 1977;25:799–803. doi:10.1021/jf60212a045

[72] Ishida M, Suyama K, Adachi S. Occurrence of dibutyl and di(2-ethylhexyl) phthalate in chicken eggs. J Agric Food Chem. 1981;29:72–74. doi:10.1021/jf00103a019

[73] Ishida M, Suyama K, Adachi S, Hoshino T. Distribution of orally administered diethylhexyl phthalate in laying hens. Poult Sci. 1982;61:262–267. doi:10.3382/ps.0610262

[74] Yang M, Yang Y, Nie S, Xie M, Chen F, Luo PG. Formation of trans fatty acids during the frying of chicken fillet in corn oil. Int J Food Sci Nutr. 2014;65:306–310. doi:10.3109/09637486.2013.858237

[75] Filazi A, Yurdakok B. Residue problems in milk after antibiotic treatment and tests used for detection of this problem. Türkiye Klinikleri J Vet Sci. 2010;1:34–43.

[76] Kan CA. Prevention and control of contaminants of industrial processes and pesticides in the poultry production chain. World Poultry Sci J. 2002;58:159–167. doi:10.1079/WPS20020015

[77] Chafer-Pericas C, Maquieira A, Puchades R. Fast screening methods to detect antibiotic residues in food samples. Trends Analyt Chem. 2010;29:1038–1049. doi:10.1016/j.trac.2010.06.004

[78] Stolker AAM, Brinkman UAT. Analytical strategies for residue analysis of veterinary drugs and growth-promoting agents in food-producing animals-a review. J Chromatogr A. 2005;1067:15–53. doi:10.1016/j.chroma.2005.02.037

[79] Pikkemaat MG. Microbial screening methods for detection of antibiotic residues in slaughter animals. Anal Bioanal Chem. 2009;395:893–905. doi:10.1007/s00216-009-2841-6

[80] Reig M, Toldra F. Veterinary drug residues in meat: Concerns and rapid methods for detection. Meat Sci. 2008;78:60–67. doi:10.1016/j.meatsci.2007.07.029

[81] De Wasch K, Okerman L, De Brabander H, Van Hoof J, Croubels S, De Backer P. Detection of residues of tetracycline antibiotics in pork and chicken meat: correlation between results of screening and confirmatory tests. Analyst. 1998;123:2737–2741. doi:10.1039/A804909B

Mycotoxins in Poultry

Ayhan Filazi, Begum Yurdakok-Dikmen,
Ozgur Kuzukiran and Ufuk Tansel Sireli

Abstract

Mycotoxins, the toxic secondary metabolites of fungi, particularly produced by many species of *Aspergillus*, *Fusarium* and *Penicillium*, have affected animal and human health for over thousand years, whereas little has been discovered so far about these complex substances in poultry, which are generally very sensitive. Even though it varies by species and sex, some common effects are reduced feed intake, weight gain, feed efficiency, growth performance, immunity and hatchability along with increased mortality, organ damages (mainly kidney and liver), carcinogenicity, teratogenicity and decreased egg production. Besides their adverse health effects and the decrease in production rate, concerns over their importance in public health is still under debate. Decontamination approaches to reduce mycotoxins in feed are technologically diverse and based on chemical, biological and physical strategies. Chemical remediation strategies involve the conversion of mycotoxins via chemical reactions. Biological strategies involve various substances such as plant ingredients, enzymes and microorganisms. Physical processes include sorting, milling, dehulling, cleaning, heating, irradiation or combinational approaches. New strategies for the prevention and treatment of mycotoxicosis, including beneficial microorganisms/products, along with alternative treatments, including plant extracts/essential oils, are current hot topics in the poultry industry.

Keywords: Control, mycotoxins, poultry, prevention

1. Introduction

Mycotoxins, the secondary metabolites of fungi, are a global concern. At aerobic conditions, fungal growth in various feed raw materials is inevitable. There are about 200 species of fungi that produce mycotoxins. Majority of the fungi that form mycotoxin belong to three genuses: *Aspergillus, Penicillium* and *Fusarium*. Although more than 500 mycotoxins produced by these fungi are known, only some of these mycotoxins exert pathogenic characteristics.

The poisoning in humans and animals caused by feeds and foods contaminated with mycotoxins may range from a slight reaction to death [1–6].

Fungal growth and mycotoxin production initiate in the cropland, during transportation or storage, and are affected by the environmental conditions including seasons, location of grain cultivation, drought and time of harvest. Long-term analyses show that feed and feedstuffs may be contaminated with mycotoxins, where these contaminated feed materials often include more than one mycotoxin [7]. Also, each of the cereals and oil seeds, available at the poultry feeds are vegetable substances obtained in different climatic conditions during vegetation in the cropland, transport and storage. For this reason, although generally only one mycotoxin is produced in raw feed materials, multiple types of mycotoxins might be found in mixed feeds. Such co-contamination examples in poultry feed are as follows: aflatoxin presence with ochratoxins, T-2 toxin or diacetoxyscirpenol; ochratoxins with T-2 toxin or citrinin and vomitoxin with fumaric acid in the poultry feeds [8].

According to the Food and Agriculture Organization Report, 25% of the world's growing crops are affected by mycotoxins each year, with annual losses of around 1 billion metric tons of food and food products [9]. Generally, there is yield loss or reduced crop value due to diseases induced by toxigenic fungi, and losses in animal productivity and animal or human health costs are due to mycotoxin contamination. Apart from these, the extra costs include the management of mycotoxin, such as prevention, control, sampling, mitigation, labor loss and research costs. Thus, the economic problems related to mycotoxins concern all sections of society [10].

The reasons for mycotoxins to receive this particular attention are their undesirable health effects, decrease in the production rate due to the spoilage of feed, and overall economic effects which are reflected in international trade of food and food products. Therefore, control of the fungal development and mycotoxin production are crucial for feed and animal producers [6].

Mycotoxins are metabolized in the alimentary canal, liver or kidneys of the poultry in accordance with their chemical properties. Their transfer to poultry meat and eggs leads to undesirable health effects in humans, leading to major concerns in public health. Contamination of the feeds with fungi both damages their organoleptic properties and increases poisoning risk by decreasing their nutritional value. Toxicity of the mycotoxins depends on the amount of absorption, number of the metabolites that are formed, exposure period and sensitivity of the animal [1].

Some mycotoxins like aflatoxins (AF), ochratoxin A (OTA), fumonisins (FUM), deoxynivalenol (DON) and T-2 toxin significantly affect the health and productivity of poultry species [11]. The aim of this review is to discuss in detail the important mycotoxins for poultry and their effects, along with the recent developments in prevention strategies.

2. Selected mycotoxins in poultry production

2.1. Aflatoxins

Aflatoxins, a group of harmful secondary metabolites characterized by polyketide-derived furanocoumarins, are produced mainly by Aspergillus fungi, such as *A. flavus* and

A. parasiticus. Aflatoxins B1 (AFB1), G1 (AFG1) and their dihydroxy derivatives B2 (AFB2) and G2 (AFG2) naturally contaminate feeds. The presence of Aflatoxin M1 and M2 (AFM1 and AFM2), the 4-hydroxy metabolites of AFB1 and B2 in biological fluids including milk and tissues, is related to the exposure of the contaminated feed. International Agency for Research in Cancer (IARC) classified these highly toxic compounds as highly carcinogenic to humans (Group 1) [12].

AFB1 was explored in the early 1960s as the main etiological agent of "Turkey X Disease" responsible for the death of young turkeys in England as a result of contaminated peanut-based feed [13]. It is a widespread dietary hepatotoxin and hepatocarcinogen, and a major public health concern throughout the world. There are substantial species-specific differences with regard to susceptibility to the toxic effects of AFB1, and domestic turkeys (*Meleagris gallopavo*) are among the most susceptible species known so far [14, 15].

Aflatoxins are usually found in feed ingredients used for poultry rations. Most extensive forms of AF include B1, B2, G1 and G2, with AFB1 being the most widespread and biologically active form [16]. In fact, AFB1 is a "pro-carcinogen" that is activated to a reactive form by the enzyme hepatic microsomal cytochrome P450 (CYP450), whereas electrophilic AFB1-8,9-epoxide (AFBO) is required for carcinogenic and toxic activity [13]. This compound forms AFB-N7-guanine adduct with DNA, which is not stable and is transformed into formamidopyrimidine. DNA adducts and repair activities through modulation are considered as important markers in carcinogenesis susceptibility. AFB-N7-guanine adduct in urine is also a potential biomarker of AFB1 exposure in animals and humans, and is vital for estimating exposure conditions and potential risk in individuals consuming AFB1 [12].

Major AFB1 detoxification route is via conjugation of the AFBO to endogenous glutathione (GSH) catalyzed by the classical detoxification enzymes glutathione S-transferases (GSTs) in mammals. Xenobiotics, including chemical carcinogens and environmental contaminants, are metabolized through detoxification processes in phase-II metabolism through these proteins [17]. Due to the expression of A3 subunit (mGSTA3), mice bioactivate AFB1 and are assumed as AFB1-resistant with great catalytic activity for AFBO. The present approach is that efficiency of GST conjugation is a major "rate-limiting" determinant for AFB1 action in individuals and species, irrespective of the efficiency of AFB1 bioactivation [14].

Aflatoxins cause a variety of effects in poultry, including decreased weight gain; poor feed efficiency; reduced egg production and egg weight; increased liver fat; changes in organ weights; reduction in serum protein levels; carcass bruising; poor pigmentation; liver damage; decreased activities of several enzymes involved in the digestion of starch, protein, lipids, and nucleic acids; and induction of immunosuppression. Evidence suggests that immunosuppression caused by AF results in many disease outbreaks, vaccination failures and poor antibody titers [9, 11]. At necropsy, livers are usually pale and enlarged, as a result of aflatoxicosis. Histologically, liver lesions include congestion of the hepatic sinusoids, focal hemorrhages, centrilobular fatty cytoplasmic vacuolation and/or necrosis, biliary hyperplasia, and nodular lymphoid infiltration. AF produces a malabsorption syndrome characterized by steatorrhea, hypocarotenoidemia, and decreased concentrations of bile salts and pancreatic lipase, trypsin, amylase and RNase at levels that do not affect growth [11].

Broiler chicken fed with 1.0 mg of AFB_1/kg of diet were found to show decreased hepatic gene expression of superoxide dismutase, GST, and epoxide hydrolase and increased gene expression of interleukin-6 and CYP1A1 and 2H1 at cellular level [18].

Ingestion of 2 ppm AFB1 in male broiler chicks was found to alter various hepatic genes causing up-down regulation. For instance, enzymes having role in the production of energy and metabolism of fatty acids (carnitine palmitoyl transferase), development and growth (insulin-like growth factor 1), coagulation (coagulation factors IX and X), protection of immune system (interleukins), antioxidant protection (GST), detoxification (epoxide hydrolase) were found to be downregulated; while cell-proliferation enzymes (ornithine decarboxylase) were upregulated [19].

A study reported that wild turkeys are significantly more resistant to AFB1 compared to domestic turkeys. Intensive breeding technologies and industrial alliance to produce modern domestic turkey led to the unintentional loss of AFB1-protective GST alleles directing a relative resistance. Actually, it has been shown that similar breeding pressures have eventuated in a remarkable loss of rare alleles and genetic diversity of single-nucleotide polymorphisms in commercial breeds of chickens [14].

As mentioned previously, mycotoxins not only lead to the aforementioned economic and health problems in poultry, but also cause public health concerns due to their residues in food for human consumption. Major metabolites of AFB1 formed in chicken liver are AFM1 and AFB2a. AFB1 and B2 are then degraded to cyclopentanol and aflatoxicol through NADP. Both AFB1 and aflatoxicol are known to accumulate at the layer of the egg. While AFB1 and AFM1 are present in chicken muscle and blood, the levels are found much higher in turkeys; the aflatoxicol levels were found to be less prominent in these animals. As a comparison, 1/1200 of AFB1 taken with feeds was found to accumulate in poultry meats, while 1/2200 of AFB1 was found to accumulate in the eggs [20].

2.2. Ochratoxins

Ochratoxins are a family of structurally related metabolites that are produced by Aspergillus and Penicillium species, including *A. ochraceus*, *A. niger* and *P. verrucosum* [21]. The most prevalent form is ochratoxin A (OTA) followed by its non-chlorinated metabolite ochratoxin B (OTB) and the ethyl ester form ochratoxin C (OTC). OTA is the most frequent and relevant form of this family, while OTB and OTC are generally counted to be of lesser importance [22]. IARC classified ochratoxin A as a compound possibly carcinogenic to humans (Group 2B) [23].

Aspergillus species can generate OTA and OTB in parallel, and experiments with *A. ochraceus* have ascertained growth-associated production of OTA and OTB, in which the yield and the ratio were dependent on the current culture terms. Mostly, the amount of OTB generated was quite lower than that of OTA, but under some situations, the level of OTB production was comparable to that of OTA. The informed generation ratios (OTA:OTB) ranged from 2:1 to 34:1 [22]. Herein, it was reported that a complex interaction of various carbon sources, basal media and nitrogen sources seems to be considerable. High OTA production was related to an induction of OTA polyketide synthase expression, whereas OTB production is not connected with transcription of the polyketide synthase gene. Laboratory fermentation experiments with *A. ochraceus* result in production of OTA at high yields (by 10 mg/g), OTB and temporarily also ochracin [24]. The intermediate metabolite OTβ was determined to be biotransformed in an effective manner into both OTA and OTB (14% and 19%, respectively), whereas OTα was biotransformed only into OTA (4.9%). In addition, OTB is inadequately converted (1.5%) into OTA, whereas some OTB may be produced by dechlorination of OTA [22].

OTA is hepatotoxic, nephrotoxic, neurotoxic, teratogenic and immunotoxic as confirmed by *in vivo* experiments with different animal species and various *in vitro* methods; its adverse effects include renal toxicity and carcinogenesis. Molecular studies with OTA revealed a non-DNA-reactive genotoxic mechanism, which includes various epigenetic mechanisms principally connected to oxidative stress, compensatory cell proliferation and disruption of cell signaling and division [25]. However, a direct genotoxic mechanism including OTA bioactivation and DNA adduct formation was also suggested [23] and this mechanism was found to be in accordance with some *in vivo* gene expression results [25]. Overall, the mode of action of OTA for renal carcinogenesis is yet under discussion.

Ochratoxins cause significant health problems and economic losses in poultry [26] and cause mycotoxic porcine nephropathy (MPN) [27]. Ochratoxin-related diseases are characterized by severe kidney damage, which could be overtly related to the exposure to ochratoxins, sometimes in combination with different mycotoxins [27]. Likewise, a slow, progressive renal disease (endemic nephropathy, EN), characterized by cellular interstitial fibrosis, tubular atrophy, and karyomegaly predominately in proximal convoluted tubules was described in humans. The etiology of this disease is still unknown, but researchers agree that the causative agent is of natural origin. The most common causes of the multiethiologic disease, EN, were the aristolochic acid from the plant birthwort (*Aristolochia clematitis*) and mycotoxins (OTA and citrinin) [28].

OTA consists of an isocoumarin moiety linked through the 7-carboxy group to the amino acid L-β-phenylalanine. OTA interferes with DNA, RNA and protein synthesis by inhibiting the enzyme phenylalanine-tRNA synthetase at a cellular level. It also affects renal carbohydrate metabolism through the reduction of the renal mRNA coding for phosphoenolpyruvate carboxykinase, a key enzyme in gluconeogenesis [11]. The effects of OTA on DNA, RNA and protein synthesis are thought to be due to the phenylalanine moiety of the toxin competing with phenylalanine in the enzyme-catalyzed reaction. OTA also causes hypocarotenidemia which has more severe effects in broilers than AF [29].

Signs of OTA toxicity in poultry include weakness, anemia, decreased feed consumption, reduced growth rate and egg production, poor feathering and excessive mortality at high dietary concentrations [21]. Pathophysiological changes include decreased urine concentration and glomerular filtration rate, impairment of proximal tubular function, and degeneration and ultrastructural alterations in renal integrity [30]. Increases in the relative weights of liver, spleen, pancreas, proventriculus, gizzard and testes have also been reported in poultry fed OTA [21]. A study found that the expression of *Eimeria tenella* and its pathological effects were maximum in the presence of OTA compared to the incidence of coccidiosis alone in broiler chicks [31].

Comparative toxicity studies of OTA and OTB have shown that *in vivo* and *in vitro* effects are very different. OTB is overtly less toxic *in vivo* compared to OTA as indicated in different models. LD50 values, found in a comparative study using 1-day-old chicks, were 120 µg for OTA (about 3.5 mg/kg) and 1890 µg for OTB (54 mg/kg) [32]. OTB is more easily excreted and has a lower affinity for plasma proteins, which may partly elucidate its lower toxicity. Both OTA and OTB toxins induce acute cytotoxic effects *in vitro*, ensuring similar amounts are taken up and are intracellularly bound, while other complex molecular mechanisms were introduced for chronic cytotoxicity studies. Moreover, it can be supposed that the small structural difference, although not responsible for the toxicity, may be crucial for the differential uptake and binding in cells. Furthermore, OTC seems to be similarly acute toxic *in vivo* and *in vitro* compared to OTA; however, the mode of action of OTC and OTA remains to be explained. In a study, oral LD50 values were reported for OTC (216 mg animal^{-1}) and OTA (166 mg animal^{-1}) in day-old chicks. Other ochratoxin ethyl or methyl esters showed lower toxicity compared to OTA. In comparison to OTA, the methyl ester of OTA was less toxic than OTA in day-old chicks, while OTB methyl and ethyl esters were found to be non-lethal to orally exposed day-old ducklings [22]. OTα is much less toxic (approximately 100 fold) than OTA as indicated in different studies [33]. It is obvious that the isocoumarin moiety alone is not effective but must be bound to phenylalanine to show toxic effects. With the current knowledge, no clear general toxicity ranking can be drawn; after all, OTA seems to be overall the most toxic, followed by OTC, OTB and OTα [33].

2.3. Fumonisins

Fumonisins (FUM) are a group of mycotoxins that were first isolated from cultures of *Fusarium moniliforme* and chemically characterized in 1988 by Gelderblom and colleagues [34]. Six different FUM have been identified (A1, A2, B1, B2, B3, B4) and their structures elucidated. However, fumonisin B1 (FB1) has been reported to be the predominant form produced by *Fusarium moniliforme*. Several other Fusarium species and a species of Alternaria have also been found to produce FB1 [35]. Based on all these animal studies, FB1 is classified by IARC as possibly carcinogenic to humans (Group 2B) [36].

The metabolism of fumonisin is yet to be elucidated. FB1 is metabolized into partially hydrolyzed FB1 and then to the hydrolyzed form (HFB1) in both gastrointestinal tract and liver, where it persists at low concentration for few more days in pigs [37]. FB1 was found to be more toxic than HFB1 in piglets [38]. Even though N-acylation of FB1 and the formation of HFB1 are shown in human cell lines and in rats, the metabolism in the avian species still remains uncertain and yet it is not possible to generalize the metabolic pathways in all animal species [39].

The mechanism that causes toxicity of fumonisins in animals seems to be due to the disruption of sphingolipid metabolism. Present evidence shows that the FUM are specific inhibitors of ceramide synthase (sphinganine/sphingosine N-acyltransferase), a key enzyme needed for the synthesis of ceramide and more complex sphingolipids. Inhibition of this enzyme system causes an increase in tissue concentrations of the sphingolipids sphingosine (SO) and sphinganine (SA), and a change in the SA:SO ratio. An increase in the SA:SO ratio has been demonstrated in tissues of broilers, turkeys, and ducklings fed FB1 [40].

In comparison to horses and swine, two susceptible species, chicks and turkeys, are relatively resistant to the toxic effects of FB1. Mild to moderate toxicity was reported in chicks, ducks and turkeys fed rations containing 75–400 mg FB1/kg for 21 days. The primary changes in chicks, ducks and turkeys were decreased body weight gain and liver pathology [41–43]. Hepatic changes in chicks were multifocal hepatic necrosis and biliary hyperplasia. Hepatocellular hyperplasia and increased extramedullary hematopoiesis were also noted in one study [44]. The primary liver pathology observed in turkeys fed with 150–300 mg FB1/kg [43] and ducklings fed with 400 mg FB1/kg [41] were diffuse hepatocellular hyperplasia, with biliary hyperplasia (more evident in turkeys). In studies designed to evaluate the chronic effects of FB1, chick performance up to 7 weeks was not affected by up to 50 mg FB1/kg diet, whereas turkeys fed with 50 mg FB1/kg diet had lower feed intakes than birds fed 0 or 25 mg FB1/kg diet [45].

2.4. Trichothecenes

Trichothecene mycotoxins are a group of fungal metabolites with the same basic backbone structure and include T-2 toxin, HT-2 toxin, diacetoxyscirpenol (DAS), monoacetoxyscirpenol (MAS), neosolaniol, 8-acetoxyneosolaniol, 4-deacetylneosolaniol, nivalenol, 4-acetoxynivalenol (Fusarenone-X), DON (vomitoxin) and 3-acetyldeoxynivalenol. They are known as the most potent small molecule inhibitors of protein synthesis and the main toxic effect at the cellular level appears to be the primary inhibition of protein synthesis followed by a secondary disruption of DNA and RNA synthesis [11]. The overall conclusion by IARC was that toxins derived from *Fusarium sporotrichioides* are not classifiable as to their carcinogenicity to humans (Group 3) [46].

For livestock, the most important trichothecene mycotoxin is DON, which is commonly a contaminant of corn, wheat and other commodity grains. Lesser amounts of T-2 toxin and DAS are found sporadically in the same sources. Poultry and cattle are more tolerant of trichothecenes than are pigs. Compared to the related DON, T-2 toxin is less frequent in crops. Some reports indicate that trichothecenes such as DON, nivalenol and fusarenon X are more frequent (57%, 16% and 10% of tested grain samples) in European grain samples than other trichothecenes like T-2 toxin (20%), HT-2 toxin (14%), T-2 tetraol (6%), neosolaniol (1%), DAS (4%), MAS (1%) [47].

Trichothecene poisoning in poultry is acute or chronic. Acute poisoning has a characteristic clinical picture and can be readily diagnosed, while chronic poisoning shows unspecific clinical symptoms [48].

Toxic effects of trichothecenes include oral lesions, growth retardation, abnormal feathering, decreased egg production and egg shell quality, regression of the bursa of Fabricius, peroxidative changes in liver, abnormal blood coagulation, leucopoenia and proteinemia, and immunosuppression [49]. Concentrations of T-2 that cause oral lesions are lower (0.4 mg/kg) than concentrations reported to decrease chick performance (3–4 mg/kg) [11]. In a comprehensive review, Danicke [49] concluded that broiler performance is affected at dietary concentrations of 3–4 mg/kg of T-2 toxin, whereas ducks were affected when the dietary concentration was as low as 0.4 mg/kg.

T2 toxin was found to decrease the immune response, represented by the decrease of lymphoid cells in the bone marrow, thymus and spleen causing resistance to infectious diseases including salmonellosis and *Escherichia coli* and cause resistance to treatments of these diseases in poultry [50]. In broilers, T-2 toxin may cause a decrease in body weight and relative weights of bursa of Fabricius, thymus, and spleen, enlarged liver, friable, and yellowish discoloration with distended gall bladder during *Mycoplasma gallisepticum* infections. Microscopical findings include vacuolar degeneration along with augmented hyperplasia in bile duct epithelia; Kupffer cell activity and infiltration of inflammatory cells in liver; vacuolar degeneration with pyknotic nuclei in kidney; lymphocytolysis and reduction of prominent reticuloepithelial cells in lymphoid organs; desquamation of villous-type epithelial cells and lymphoid intrusion in the submucosa of proventriculus; mild hemorrhage along with inflammatory cells in the heart; desquamation and erosion of the mucosa in trachea and the thickening of the air sacs along with edema and the presence of inflammatory cells in air sacs [51]. The toxic effect also manifests as reduced proliferation of lymphocytes stimulated by phytohemagglutinin and lipopolysaccharide in Pekin duck broilers [52].

DON was found to be less toxic than T-2 toxin, and the level of DON that affects chick performance is still disputed. Some researchers [53, 54] reported toxic effects at 16 mg/kg diet, whereas others [55] report no toxic effect until dietary concentrations exceeded 116 mg/kg of DON. A review paper summarizing results of 49 studies with DON concluded that a dietary concentration of 5 mg/kg had no negative effects on performance [56]. DON has also been demonstrated to have both immunosuppressive and immunomodulating effects in poultry [49]. Recent studies indicate that DON at concentrations ranging from 1 to 7 mg/kg diet significantly alters several key functions of the intestinal tract including decaying villus surface area available for absorption and altering the permeability of the alimentary canal [57].

3. Interactions among mycotoxins

In nature, co-occurrence of mycotoxins is generally observed. Meanwhile, for many years the research community focused on the occurrence of singular mycotoxins. Nowadays, scientific interest is shifted to studies involving multiple mycotoxins using various co-occurrence scenarios. One fungus may produce many different mycotoxins, and the same mycotoxin may be produced by several species. A paper conducted a meta-analysis of publications (> 100) describing toxicological interactions among mycotoxins. Results indicated that most of the studies showed a synergistic or additive interaction on animal performance. However, results with respect to other response variables indicated that there were many types of interactions ranging from synergistic to antagonistic for the same association [58]. They also observed from their review that a combination of mycotoxins, at concentrations that individually should not cause negative effects, may negatively affect animals.

The individual and combined effects of dietary AFB1 and FB1 on liver pathology, serum levels of aspartate amino-transferase (AST) and plasma total protein (TP) of broilers were quantified from 8 to 41 days of age with the dietary treatments of AFB1 (0, 50 and 200 μg AFB1/kg), and FB1 (0, 50 and 200 mg FB1/kg). Following treatment, AST levels were found to be higher in all treatment groups (except 50 mg FB1) compared to controls at day 33. TP levels were found to be reduced at 6 days post feeding in AFB1-treated group (200 μg) and in FB1 combination group. At 33 days post feeding, the combination group (200 μg AFB1 and 200 mg FB1) were found to have higher plasma TP, proliferation of bile duct and trabecular disorders in liver tissue compared to control; while the changes in other groups were insignificant compared to controls. Overall AFB1 alone and in combination caused damage in liver at varying degrees and an increase of serum AST levels [59].

Aflatoxicosis causes a reduction in the production of egg and a decrease in egg weight in laying hens. Meanwhile, the antagonistic effects of AF and FB on egg production were reported in quails, where the decrease in egg production in FB-only treated group was much evident compared to AF+FB combination [60].

Few studies examined the combined effect of AF and FB on immunity. AF and FB co-contaminated feed was found to reduce lymphocyte proliferation by mitogenic stimulation as less than additive [61] or as additive [62] compared to single contamination. A study indicated a synergistic decrease of the antibody titers against Newcastle disease [59]. On the contrary, another study demonstrated an unexpected increase and an additive effect of the two toxins when looking at the hemagglutination titers against sheep red blood cells in turkey poults. However, the phytohemagglutinin delayed hypersensitivity response was not affected by dietary treatment. These results indicate that FB1 and AF, alone and in combination, can adversely affect poult performance and health [63].

As is known, both AF and OTA reduce egg production and hatchability. The combined effects of these two toxins were studied in laying hens [64, 65]. An additive interaction of AF and OTA was observed on egg production and on the feed efficiency (consumption for egg production) [64]; meanwhile this interaction was dependent on the concentration, resulting from synergistic to slightly lower than additive effect and also modulates the protein and energy usage [65].

AF- and OTA-contaminated feed resulted in microscopic lesions in the liver and kidneys, along with respective target organs in chicken, while contradictory results are presented in different studies. As such, OTA in the diet was found to prevent the hepatic fatty infiltration caused by AF in chicken [66]. Pigs fed the co-contaminated diet offer the same hepatic lesions as those fed with the diet contaminated with AF alone [67]. On the contrary, a study recorded more severe hepatic lesions in chickens taking the co-contaminated diet, with granular and vacuolar degenerative changes, necrosis of liver parenchyma and areas of hemorrhages [68]. The same conflict was realized for the histology of the kidney. In pigs, less severe renal lesions and lower creatinine and blood urea nitrogen concentrations were observed in animals fed the co-contaminated diet compared to animals fed the OTA-contaminated diet [67]. In contrast, a study observed that renal injuries appeared earlier and were more developed in chickens fed a multi-contaminated diet than in animals taking the mono-contaminated diets, which caused destruction of tubular epithelium, with detachment of tubular cells from basement membrane [68]. The species used may explain these conflicts. Apart from that, chronic DON exposure did not induce any effect on FB1 toxicokinetics in broilers [69].

The results on combination toxicity are yet quite limited and occasionally conflicting. Nowadays, very little is known about mycotoxin interactions although combined exposure is clearly more relevant to real-life conditions. It is known that the combined effects of mycotoxins are mostly additive or synergistic; whereas depending on the concentrations and the *in vitro* model employed, antagonistic interactions have also been determined. The results on multiple mixtures are still inadequate [22].

4. Prevention and control

Prevention of fungi production in the feeds may be achieved by always keeping the feeds fresh, keeping the humidity low and equipment clean and also adding fungistatic substances. Humidity exceeding 11% promotes fungal production in cereals and feed. Storage conditions that afford high relative humidity also significantly affect humidity content of the feed. Good ventilation of the storehouse removes humidity from the raw material of the feed and storehouse. Physically damaged cereals are more prone to fungus production compared to the healthy ones. Changing the raw materials at the places where they are stored at short intervals decreases mycotoxin formation [2, 3, 6].

Research efforts progressively increase to develop mitigation strategies based on risk monitoring, risk characterization, prevention, intervention, and remediation strategies for multiple mycotoxins, initiating from critical points along the production chain comprising field, storage, processing and transportation. However, monitoring and good agricultural, storage, and transportation practices along with an effective Hazard Analysis and Critical Control Point (HACCP) approach do not completely prevent mycotoxin presence in the food or feed chain. Decontamination strategies then offer a last resort to salvage contaminated batches along the production chain [70].

Considering the variation of mycotoxin structures, it could be inferred that there is no single method which can be used to deactivate mycotoxins in feed. Therefore, different strategies have to be combined in order to specifically target individual mycotoxins without affecting the quality of feed [11].

Decontamination strategies to reduce mycotoxins in food and feed commodities are technologically diverse and based on chemical, biological and physical approaches. Chemical remediation strategies involve the conversion of mycotoxins via chemical reactions. Ammoniation, alkaline hydrolysis, peroxidation, ozonation and the use of bisulphites are reported to be effective on one or more mycotoxins but a detailed insight into the toxicity of eventual end products or the impact on palatability and nutritive quality is questionable [71].

Biological approach in treatment strategies involves various substances (algae, plant ingredients, etc.) that protect critical organs such as the liver and strengthen the immune system of animals. Enzymatic or microbial detoxification, also referred to as "biotransformation" or "biodetoxification", uses microorganisms or purified enzymes thereof to catabolize the entire mycotoxin or transform or cleave it to less or non-toxic compounds [11]. Some microorganism such as *Rhodococcus erythropolis* [72], *Armillariella tabescens* [73] and *Myxococcus fulvus* [16] have been suggested to have different AF-degrading ability. *Rhizopus oryzae* [74], *Bacillus licheniformis* [75] and Pseudomonas sp [76] were searched for their abilities to degrade ZEA. Some studies found that *B. subtilis* had protective effects against aflatoxicosis in layers and broilers fed naturally AF-contaminated diets and also healed ZEA toxicosis in pre-pubertal gilts when fed diets including ZEA. Therefore, *B. subtilis*, as a new feed additive for biodegradation of AF and ZEA may have promising potential in feed industrial applications [77].

Some physical processes aim to remove highly contaminated fractions from bulk material through sorting, milling, dehulling, cleaning, heating, irradiation or combinational approaches [78, 79]. Another physical removal strategy is the use of inorganic or organic mycotoxin binders [80]. Due to low feed inclusion requirements and easy management of AF enterosorbents, the widespread acceptance of these products by the farm animal industry has led to the introduction of a variety of diverse materials and/or complex mixtures for AF binding. These have been labeled as mycotoxin enterosorbents, binders, sequestrants, interceptor molecules, trapping agents, adsorbents, toxin sorbents, and so on. These materials (and/or mixtures) are reported to contain smectite clays, zeolites, kaolinite, mica, silica, charcoal, sodium bentonite and various biological constituents including chlorophyllins, yeast products, lactic acid bacteria, plant extracts and algae. Some contain smectite or zeolite minerals that have been amended with natural or synthetic surfactants resulting in hydrophobic organoclays or organozeolites [81–84]. There is considerable evidence indicating that smectite clays are the most effective AF enterosorbents. Although these adsorbing binders have some promising features, some may have adverse nutritional effects due to binding of vitamins and minerals or reducing the efficacy pharmacokinetics of antibiotics [85]. Also, possible dioxin contamination might pose a risk for using natural clays in case of forest and trash fire near the sources [86]. Furthermore, the adsorption efficacy of binding agents is limited to only a few mycotoxins, such as AF, ergot alkaloids, and some other fungal toxins, while binders have been shown to be ineffective for trichothecenes [87]. Therefore, alternative approaches for efficient detoxification of mycotoxins are required.

Use of microorganisms and their specific products such as enzymes to detoxify specific mycotoxins not only work for non-adsorbable mycotoxins, but for all other toxins for which respective microbes can be isolated from nature. This approach has been known for a long time, even longer than the binder concept. Within few years after the discovery of AF, the first report on a bacterium capable of detoxifying AF by catabolization was published [88]. Since then, many microorganisms were isolated from different habitats such as the gastrointestinal tract (GIT) of animals, soil, mycotoxin-contaminated materials (e.g., grains) and insects feeding on such materials. The ability of various bacteria, yeast, fungi and enzymes in detoxifying mycotoxins by transformation, cleavage and catabolization has been recently reviewed [89]. However, only a few of these organisms were useful or further investigated for practical applications in animal nutrition. Such microorganisms or enzymes need to fulfill many different requirements before they can be used for gastrointestinal detoxification of mycotoxin in animals, such as:

- The microorganism and its reaction products need to be non-toxic and safe.

- High detoxification reactivity.

- Good technological properties (fermentation, downstream processing, stabilization).

- High stability in feed and during feed processing.
- No negative impact on feed (ingredients).
- Compatibility and stability in the GIT.
- Detoxification reaction in the GIT needs to be fast and as complete as possible.

One of the microorganisms which has been further developed for practical application is *Trichosporon mycotoxinivorans*, a yeast strain capable of detoxifying OTA and ZEN [90]. Application of this yeast in poultry diets has been proven to detoxify OTA. Another organism is an anaerobic rumen bacterium BBSH 797 (Genus Novus of family Coriobacteriacae, formerly Eubacterium) which was isolated and developed as a trichothecene-detoxifying feed additive [91]. BBSH 797 detoxifies trichothecenes by cleavage of the 12, 13 epoxide ring resulting in deepoxy trichothecenes. Several microorganisms, mainly aerobic bacteria and also yeasts, with FUM degradation properties were also explored and isolated in order to detoxify FUM. However, for various reasons, none of these microorganisms were found to be useful as a mycotoxin-deactivating feed additive. Therefore, the catabolic pathway of FUM degradation was investigated and the gene coding for the key enzyme of FUM detoxification (FUMzyme) was identified, cloned and expressed in a yeast strain [92]. FUMzyme (carboxyl-esterase) was further developed and tested in swine for gastrointestinal detoxification of FUM by cleaving the tricarballylic side chains of FUM leading to the non-toxic metabolite hydrolyzed FUM (HFB1) [93].

One of the common approaches to overcome mycotoxicosis in poultry is using herbal products including essential oils as plant-based fumigants in feed storage [94]. Essential oils are complex compounds, and their chemical composition and concentrations of various compounds are variable. Essential oils basically consist of two classes of compounds, the terpenes and phenylpropenes, depending on the number of 5-carbon building blocks. For example, 500 ppm of the ethanolic extract of *Thymus vulgaris* could partially restore the negative impact of AFB1 (600 ppb) in commercial broilers [9]. They suggested that this herb can be used as natural non-antibiotic feed additive on broilers in the prevention of aflatoxicosis. As a result of the change in diet (change in nutrients, phytochemicals, contamination, xenobiotics), the levels of the drug-metabolizing enzymes (phase-I and phase-II) are expected to change, which would eventually lead to a change in AFB1 adducts. On the other hand, as phenolic phytochemicals have antioxidant effects at varying degrees due to their various chemical structures, they are assumed to have a protective role in the cellular components against free radical–induced damage caused by aflatoxicosis [95]. Apart from that, a herbal mycotoxin binder comprising of a combination of minerals (extra purified clay containing diatomaceous earth minerals), antioxidants (curcuminoids extracted from turmeric) and enzymes (Epoxidases and Esterases) in proportions of 15, 10 and 75%, respectively, partially restored feed consumption and egg production, alleviating some side effects of AFB1 (500 ppb in feed) in broiler breeders [96].

5. Conclusion

Understanding the occurrence and prevalence of mycotoxins and their individual as well as additive negative effects on poultry has become imperative. New insights on actual microbial detoxification routes are needed in the field, which could be based on the biodegradation metabolisms of non-mycotoxins found in diverse microbial communities. Indeed, many hazardous, undesirable, deleterious or recalcitrant molecules in other research fields share structural analogies with diverse mycotoxins and are reported to be successfully degraded by microorganisms. These unexplored worlds may serve as resource for cutting-edge research in the field of mycotoxin remediation or in the field of metagenomics screening surveys in search for new microbial degraders of mycotoxins. The usage of latest analytical techniques such as liquid chromatography tandem-mass spectrometry will increase the precision in determination of the concentrations of multiple mycotoxins present in agricultural commodities, at once. Latest enzymatic deactivation technologies help to eliminate the mycotoxins that cannot be bound using binder products. Overall, mycotoxins still impose a great risk for the poultry sector and alternative approaches for the prevention are still being sought by researches around the world.

Author details

Ayhan Filazi[1*], Begum Yurdakok-Dikmen[1], Ozgur Kuzukiran[2] and Ufuk Tansel Sireli[3]

*Address all correspondence to: filazi@veterinary.ankara.edu.tr

1 Department of Pharmacology and Toxicology, Faculty of Veterinary Medicine, Ankara University, Ankara, Turkey

2 Veterinary Control Central Research Institute, Ankara, Turkey

3 Department of Food Hygiene and Control, Faculty of Veterinary Medicine, Ankara University, Ankara, Turkey

References

[1] Basalan M, Hismiogullari SE, Hismiogullari AA, Filazi A. Fungi and aflatoxin B1 in horse and dog feeds in Western Turkey. Revue Med Vet. 2004;156:248–252.

[2] Becer UK, Filazi A. Aflatoxins, nitrates and nitrites analysis in the commercial cat and dog foods. Fresen Environ Bull. 2010;18:2523–2527.

[3] Demircioglu S, Filazi A. Detection of aflatoxin levels in red pepper produced in Turkey. Vet Hekim Der Derg. 2010;81:63–66.

[4] Filazi A, Ince S, Temamogullari F. Survey of the occurrence of aflatoxin M1 in cheeses produced by dairy ewe's milk in Urfa city, Turkey. Ankara Univ Vet Fak Derg. 2010;57:197–199.

[5] Gundinc U, Filazi A. Detection of aflatoxin M1 concentrations in UHT milk consumed in Turkey markets by ELISA. Pak J Biol Sci. 2009;12:653–656.

[6] Kaya S. Mycotoxins. In: Kaya S, editor. Veterinary Toxicology, 3rd ed. Ankara-Turkey: Medisan Publisher; 2014, pp. 393–433.

[7] Streit E, Schatzmayr G, Tassis P, Tzika E, Marin D, Taranu I, Tabuc C, Nicolau A, Aprodu I, Puel O, Oswald IP. Current situation of mycotoxin contamination and co-occurrence in animal feed--focus on Europe. Toxins. 2012;4:788–809. doi:10.3390/toxins4100788.

[8] Gentles A, Smith EE, Kubena LF, Duffus E, Johnson P, Thompson J, Harvey RB, Edrington TS. Toxicological evaluations of cyclopiazonic acid and ochratoxin A in broilers. Poult Sci. 1999;78:1380–1384. doi:10.1093/ps/78.10.1380.

[9] Manafi M, Hedayati M, Yari M. Aflatoxicosis and herbal detoxification: the effectiveness of thyme essence on performance parameters and antibody titers of commercial broilers fed aflatoxin B1. Res Zool. 2014;4:43–50. doi:10.5923/j.zoology.20140402.02.

[10] Filazi A, Sireli UT. Occurrence of aflatoxins in food. In: Razzaghi-Abyaneh M, editor. Aflatoxins-Recent Advances and Future Prospects. Rijeka-Croatia; InTech; 2013, pp.143–170. DOI:10.5772/51031.

[11] Murugesan GR, Ledoux DR, Naehrer K, Berthiller F, Applegate TJ, Grenier B, Phillips TD, Schatzmayr G. Prevalence and effects of mycotoxins on poultry health and performance, and recent development in mycotoxin counteracting strategies. Poult Sci. 2015;94: 1298–1315. doi:10.3382/ps/pev075.

[12] Dohnal V, Wu Q, Kuca K. Metabolism of aflatoxins: key enzymes and interindividual as well as interspecies differences. Arch Toxicol. 2014;88:1635–1644. doi:10.1007/s00204-014-1312-9.

[13] Rawal S, Kim JE, Coulombe R. Aflatoxin B1 in poultry: toxicology, metabolism and prevention. Res Vet Sci. 2010;89:325–331.

[14] Kim JE, Bunderson BR, Croasdell A, Reed KM, Coulombe RA. Alpha-class glutathione S-transferases in wild Turkeys (*Meleagris gallopavo*): characterization and role in resistance to the carcinogenic mycotoxin aflatoxin B1. PLoS One. 2013;8: e60662. doi:10.1371/journal.pone.0060662.

[15] Rawal S, Coulombe RA. Metabolism of aflatoxin B1 in Turkey liver microsomes: the relative roles of cytochromes P450 1A5 and 3A37. Toxicol Appl Pharmacol. 2011;254: 349–354. doi:10.1016/j.taap.2011.05.010.

[16] Zhao LH, Guan S, Gao X, Ma QG, Lei YP, Bai XM, Ji C. Preparation, purification and characteristics of an aflatoxin degradation enzyme from Myxococcus fulvus ANSM068. J Appl Microbiol. 2011;110:147–155. doi:10.1111/j.1365-2672.2010.04867.x.

[17] Hayes JD, Flanagan JU, Jowsey IR. Glutathione transferases. Annu Rev Pharmacol Toxicol. 2005;45:51–88. doi:10.1146/annurev.pharmtox.45.120403.095857.

[18] Yarru LP, Settivari RS, Gowda NKS, Antoniou E, Ledoux DR, Rottinghaus GE. Effects of turmeric (*Curcuma longa*) on the expression of hepatic genes associated with biotransformation, antioxidant, and immune systems in broiler chicks fed aflatoxin. Poult Sci. 2009;88:2620–2627. doi:10.3382/ps.2009-00204.

[19] Yarru LP, Settivari RS, Antoniou E, Ledoux DR, Rottinghaus GE. Toxicological and gene expression analysis of the impact of aflatoxin B1 on hepatic function of male broiler chicks. Poult Sci. 2009;88:360–371. doi:10.3382/ps.2008-00258.

[20] Hossain SA, Haque N, Kumar M, Sontakke UB, Tyagi AK. Mycotoxin residues in poultry product: their effect on human health and control. Wayamba J Anim Sci. 2011; ISSN: 2012-578X: P92–P96.

[21] Gibson RM, Bailey CA, Kubena LF, Huff WE, Harvey RB. Ochratoxin A and dietary protein. 1. Effects on body weight, feed conversion, relative organ weight, and mortality in three-week-old broilers. Poult Sci. 1989;68:1658–1663. doi:10.3382/ps.0681658.

[22] Heussner AH, Bingle LE. Comparative ochratoxin toxicity: a review of the available data. Toxins. 2015;7:4253–4282. doi:10.3390/toxins7104253.

[23] Pfohl-Leszkowicz A, Manderville RA. An update on direct genotoxicity as a molecular mechanism of ochratoxin A carcinogenicity. Chem Res Toxicol. 2012;25:252–262. doi:10.1021/tx200430f.

[24] Harris J, Mantle P. Biosynthesis of ochratoxins by *Aspergillus ochraceus*. Phytochemistry. 2001;58:709–716. doi:10.1016/S0031-9422(01)00316-8.

[25] Vettorazzi A, van Delft J, López de Cerain A. A review on ochratoxin A transcriptomic studies. Food Chem Toxicol. 2013;59:766–783. doi:10.1016/j.fct.2013.05.043.

[26] Elaroussi MA, Mohamed FR, el Barkouky EM, Atta A, Abdou AM, Hatab MH. Experimental ochratoxicosis in broiler chickens. Avian Pathol. 2006;35:263–269. doi:10.1080/03079450600817115.

[27] Stoev S, Denev S. Porcine/chicken or human nephropathy as the result of joint mycotoxins interaction. Toxins. 2013;5:1503–1530. doi:10.3390/toxins5091503.

[28] Pepeljnjak S, Klarić M. "Suspects" in etiology of endemic nephropathy: aristolochic acid versus mycotoxins. Toxins. 2010;2:1414–1427. doi:10.3390/toxins2061414.

[29] Schaeffer JL, Tyczkowski JJ, Hamilton PB. Alterations in carotenoid metabolism during ochratoxicosis in young broiler chickens. Poult Sci. 1987;66:318–324.

[30] Glahn RP, Shapiro RS, Vena VE, Wideman RF, Huff WE. Effects of chronic ochratoxin A and citrinin toxicosis on kidney function of single comb white leghorn pullets. Poult Sci. 1989;68:1205–1212.

[31] Manafi M, Mohan K, Noor Ali M. Effect of ochratoxin A on coccidiosis-challenged broiler chicks. World Mycotoxin J. 2011;4:177–181. doi:10.3920/WMJ2010.1234.

[32] Peckham J, Doupnik BJ, Jones OJ. Acute toxicity of ochratoxins A and B in chicks. Appl Microbiol. 1971;21:492–494.

[33] Xiao H, Madhyastha S, Marquardt RR, Li S, Vodela JK, Frohlich AA, Kemppainen BW. Toxicity of ochratoxin A, its opened lactone form and several of its analogs: structure-activity relationships. Toxicol Appl Pharmacol. 1996;137:182–192. doi:10.1006/taap.1996.0071.

[34] Gelderblom WC, Jaskiewicz K, Marasas WF, Thiel PG, Horak RM, Vleggaar R, Kriek NP. Fumonisins- novel mycotoxins with cancer-promoting activity produced by *Fusarium moniliforme*. Appl Environ Microbiol. 1988;54:1806–1811.

[35] Chen J, Mirocha CJ, Xie W, Hogge L, Olson D. Production of the mycotoxin fumonisin B(1) by Alternaria alternata f. sp. Lycopersici. Appl Environ Microbiol. 1992;58:3928–3931.

[36] Domijan AM. Fumonisin B(1): a neurotoxic mycotoxin. Arh Hig Rada Toksikol. 2012;63:531–544. doi:10.2478/10004-1254-63-2012-2239.

[37] Fodor J, Balogh K, Weber M, Miklós M, Kametler L, Pósa R, Mamet R, Bauer J, Horn P, Kovács F, Kovács M. Absorption, distribution and elimination of fumonisin B(1) metabolites in weaned piglets. Food Addit Contam Part A Chem Anal Control Expo Risk Assess. 2008;25:88–96. doi:10.1080/02652030701546180.

[38] Grenier B, Bracarense AP, Schwartz HE, Trumel C, Cossalter AM, Schatzmayr G, Kolf-Clauw M, Moll WD, Oswald IP. The low intestinal and hepatic toxicity of hydrolyzed fumonisin B1 correlates with its inability to alter the metabolism of sphingolipids. Biochem Pharmacol. 2012;83:1465–1473. doi:10.1016/j.bcp.2012.02.007.

[39] Guerre P. Fusariotoxins in avian species: Toxicokinetics, metabolism and persistence in tissues. Toxins. 2015;7:2289–2305. doi:10.3390/toxins7062289.

[40] Tran ST, Auvergne A, Benard G, Bailly JD, Tardieu D, Babile R, Guerre P. Chronic effects of fumonisin B1 on ducks. Poult Sci. 2005;84:22–28. doi:10.1093/ps/84.1.22.

[41] Bermudez AJ, Ledoux DR, Rottinghaus GE. Effects of *Fusarium moniliforme* culture material containing known levels of fumonisin B1 in ducklings. Avian Dis. 1995;39:879–886. doi:10.2307/1592427.

[42] Brown TP, Rottinghaus GE, Williams ME. Fumonisin mycotoxicosis in broilers: performance and pathology. Avian Dis. 1992;36:450–454. doi:10.2307/1591528.

[43] Weibking TS, Ledoux DR, Bermudez AJ, Turk JR, Rottinghaus GE. Effects on turkey poults of feeding *Fusarium moniliforme* M-1325 culture material grown under different environmental conditions. Avian Dis. 1995;39:32–38. doi:10.2307/1591979.

[44] Weibking TS, Ledoux DR, Bermudez AJ, Turk JR, Rottinghaus GE, Wang E, Merrill AH. Effects of feeding *Fusarium moniliforme* culture material, containing known levels of fumonisin B1, on the young broiler chick. Poult Sci. 1993;72:456–466. doi:10.3382/ps.0720456.

[45] Broomhead JN, Ledoux DR, Bermudez AJ, Rottinghaus GE. Chronic effects of fumonisin B1 in broilers and turkeys fed dietary treatments to market age. Poult Sci. 2002;81:56–61. doi:10.1093/ps/81.1.56.

[46] International Agency for Research on Cancer. WHO IARC monographs on the evaluation of carcinogenic risks to humans. Some naturally occuring substances: food items and constituents, heterocyclic aromatic amines and mycotoxins. Toxins derived from *Fusarium graminearum*, *F. culmorum* and *F. crookwellense*: zearalenone, deoxynivalenol, nivalenol and fusarenone X, vol. 56. IARC, Lyon; 1993, pp. 397–444.

[47] Sokolovij M, Garaj-Vrhovac V, Simpraga B. T-2 toxin: incidence and toxicity in poultry. Arh Hig Rada Toksikol. 2008;59:43–52. doi:10.2478/10004-1254-59-2008-1843.

[48] Resanovic RM, Nešic KD, Nesic VD, Palic TD, Jacevic VM. Mycotoxins in poultry production. Proc Nat Sci. Matica Srpska Novi Sad. 2009;116:7–14.

[49] Danicke S. Prevention and control of mycotoxins in the poultry production chain: a European view. World Poult Sci J. 2002;58:451–474. doi:10.1079/WPS20020033.

[50] Boonchuvit B, Hamilton PB, Burmeister HR. Interaction of T-2 toxin with Salmonella infections of chicken. Poult Sci. 1975;54:1693–1696. doi:10.3382/ps.0541693.

[51] Manafi M, Pirany N, Noor Ali M, Hedayati M, Khalaji S, Yari M. Experimental pathology of T-2 toxicosis and mycoplasma infection on performance and hepatic functions of broiler chickens. Poult Sci. 2015;94:1483–1492. doi:10.3382/ps/pev115.

[52] Rafai P, Pettersson H, Bata A, Papp Z, Glávits R, Tuboly S, Ványi A, Soós P. Effect of dietary T-2 fusariotoxin concentrations on the health and production of white Pekin duck broilers. Poult Sci. 2000;79:1548–1556. doi:10.1093/ps/79.11.1548.

[53] Huff WE, Kubena LF, Harvey RB, Hagler WM, Swanson SP, Phillips TD, Creger CR. Individual and combined effects of aflatoxin and deoxynivalenol (DON, vomitoxin) in broiler chickens. Poult Sci. 1986;65:1291–1298. doi:10.3382/ps.0651291.

[54] Kubena LF, Huff WE, Harvey RB, Phillips TD, Rottinghaus GE. Individual and combined toxicity of deoxynivalenol and T-2 toxin in broiler chicks. Poult Sci. 1989;68: 622–626. doi:10.3382/ps.0680622.

[55] Moran ET, Hunter B, Ferket P, Young LG, McGirr LG. High tolerance of broilers to vomitoxin from corn infected with *Fusarium graminearum*. Poult Sci. 1982;61:1828–1831. doi:10.3382/ps.0611828.

[56] Danicke S, Gareis M, Bauer J. Orientation values for critical concentrations of deoxynivalenol and zearalenone in diets for pigs, ruminants and gallinaceous poultry. Proc Soc Nutr Physiol. 2001;10:171–174.

[57] Osselaere A, Devreese M, Goossens J, Vandenbroucke V, de Baere S, de Backer P, Croubels S. Toxicokinetic study and absolute oral bioavailability of deoxynivalenol, T-2 toxin and zearalenone in broiler chickens. Food Chem Toxicol. 2013;51:350–355. doi:10.1016/j.fct.2012.10.006.

[58] Grenier B, Oswald IP. Mycotoxin co-contamination of food and feed: meta-analysis of publications describing toxicological interactions. World Mycotoxin J. 2011;4:285–313. doi:10.3920/WMJ2011.1281.

[59] Tessari EN, Kobashigawa E, Cardoso AL, Ledoux DR, Rottinghaus GE, Oliveira CA. Effects of aflatoxin B1 and fumonisin B1 on blood biochemical parameters in broilers. Toxins. 2010;2:453–460. doi:10.3390/toxins2040453.

[60] Ogido R, Oliveira CAF, Ledoux DR, Rottinghaus GE, Correa B, Butkeraitis P, Reis TA, Goncales E, Albuquerque R. Effects of prolonged administration of aflatoxin B1 and fumonisin B1 in laying Japanese quail. Poult Sci. 2004;83:1953–1958. doi:10.1093/ps/83.12.1953.

[61] Kubena LF, Edrington TS, Kampsholtzapple C, Harvey RB, Elissalde MH, Rottinghaus GE. Effects of feeding fumonisin B1 present in *Fusarium moniliforme* culture material and aflatoxin singly and in combination to turkey poults. Poult Sci. 1995;74:1295–1303. doi:10.3382/ps.0741295.

[62] Harvey RB, Edrington TS, Kubena LF, Elissalde MH, Rottinghaus GE. Influence of aflatoxin and fumonisin B1-containing culture material on growing barrows. Am J Vet Res. 1995;56:1668–1672.

[63] Weibking TS, Ledoux DR, Bermudez AJ, Rottinghaus GE. Individual and combined effects of feeding *Fusarium moniliforme* culture material, containing known levels of fumonisin B1, and aflatoxin B1 in the young turkey poult. Poult Sci. 1994;73:1517–1525. doi:10.3382/ps.0731517.

[64] Verma J, Johri TS, Swain BK. Effect of varying levels of aflatoxin, ochratoxin and their combinations on the performance and egg quality characteristics in laying hens. Asian-Australas J Anim Sci. 2003;16:1015–1019. doi:10.5713/ajas.2003.1015.

[65] Verma J, Johri TS, Swain BK. Effect of aflatoxin, ochratoxin and their combination on protein and energy utilization in white leghorn laying hens. J Sci Food Agric. 2007;87: 760–764. doi:10.1002/jsfa.2655.

[66] Huff WE, Doerr JA. Synergism between aflatoxin and ochratoxin A in broiler chickens. Poult Sci. 1981;60:550–555. doi:10.3382/ps.0600550.

[67] Tapia MO, Seawright AA. Experimental combined aflatoxin B1 and ochratoxin A intoxication in pigs. Aust Vet J. 1985;62:33–37. doi:10.1111/j.1751-0813.1985.tb14229.x.

[68] Sakhare PS, Harne SD, Kalorey DR, Warke SR, Bhandarkar AG, Kurkure NV. Effect of Toxiroak® polyherbal feed supplement during induced aflatoxicosis, ochratoxicosis and combined mycotoxicoses in broilers. Vet Arhiv. 2007;2:129–146.

[69] Antonissen G, Devreese M, Immerseel FV, De Baere S, Hessenberger S, Martel A, Croubels S. Chronic exposure to Deoxynivalenol has no influence on the oral bioavailability of Fumonisin B1 in broiler chickens. Toxins. 2015;7:560–571. doi:10.3390/toxins7020560.

[70] Bhat R, Rai RV, Karim AA. Mycotoxins in food and feed: present status and future concerns. Compr Rev Food Sci Food Safety. 2010;9:57–81. doi:10.1111/j.1541-4337.2009.00094.x.

[71] Vanhoutte I, Audenaert K, De Gelder L. Biodegradation of mycotoxins: tales from known and unexplored worlds. Front Microbiol. 2016;7:561. doi:10.3389/fmicb.2016.00561.

[72] Alberts JF, Engelbrecht Y, Steyn PS, Holzapfel W, van Zyl W. Biological degradation of aflatoxin B-1 by Rhodococcus erythropolis cultures. Int J Food Microbiol. 2006;109: 121–126. doi:10.1016/j.ijfoodmicro.2006.01.019.

[73] Cao J, Zhang H, Yang Q, Ren R. Efficacy of Pichia caribbica in controlling blue mold rot and patulin degradation in apples. Int J Food Microbiol. 2013;162:167–173. doi:10.1016/j.ijfoodmicro.2013.01.007.

[74] Varga J, Peteri Z, Tabori K, Teren J, Vagvolgyi C. Degradation of ochratoxin A and other mycotoxins by Rhizopus isolates. Int J Food Microbiol. 2005;99:321–328. doi:10.1016/j.ijfoodmicro.2004.10.034.

[75] Yi PJ, Pai CK, Liu JR. Isolation and characterization of a *Bacillus licheniformis* strain capable of degrading zearalenone. World J Microb Biot. 2011;27:1035–1043.

[76] Altalhi AD, El-Deeb B. Localization of zearalenone detoxification gene(s) in pZEA-1 plasmid of *Pseudomonas putida* ZEA-1 and expressed in *Escherichia coli*. J Hazard Mater. 161:1166–1172.

[77] Jia R, Ma Q, Fan Y, Ji C, Zhang J, Liu T, Zhao L. The toxic effects of combined aflatoxins and zearalenone in naturally contaminated diets on laying performance, egg quality and mycotoxins residues in eggs of layers and the protective effect of Bacillus subtilis biodegradation product. Food Chem Toxicol. 2016;90:142–150. doi:10.1016/j.fct.2016.02.010.

[78] Kaushik G. Effect of processing on mycotoxin content in grains. Crit Rev Food Sci Nutr. 2015;55:1672–1683. doi:10.1080/10408398.2012.701254.

[79] Matumba L, Van Poucke C, Njumbe Ediage E, Jacobs B, De Saeger S. Effectiveness of hand sorting, flotation/washing, dehulling and combinations thereof on the decontamination of mycotoxin-contaminated white maize. Food Addit Contam Part A Chem Anal Control Expo Risk Assess. 2015;32:960–969. doi:10.1080/19440049.2015.1029535.

[80] Kolosova A, Stroka J. Substances for reduction of the contamination of feed by mycotoxins: a review. World Mycotoxin J. 2001;4:225–256. doi:10.3920/WMJ2011.1288.

[81] Eraslan G, Akdogan M, Arsan E, Essiz D, Sahindokuyucu F, Hismiogullari SE, Altintas L. Effects of aflatoxin and sodium bentonite administered in feed alone or combined on lipid peroxidation in the liver and kidneys of broilers. Bull Vet Inst Pulawy. 2004;48:301–304.

[82] Manafi M, Umakantha B, Swamy HDN, Mohan K. Evaluation of high-grade sodium bentonite on performance and immune status of broilers, fed ochratoxin and aflatoxin. World Mycotoxin J. 2009;2:435–440. doi:10.3920/WMJ2009.1136.

[83] Manafi M. Counteracting effect of high grade sodium bentonite during aflatoxicosis in broilers. J Agric Sci Tech. 2012;14:539–547.

[84] Ortatatli M, Oguz, H. Ameliorative effects of dietary clinoptilolite on pathological changes in broiler chickens during aflatoxicosis. Res Vet Sci. 2001;71:59–66. doi:10.1053/rvsc.2001.0487.

[85] De Mil T, Devreese M, Broekaert N, Fraeyman S, DeBacker P, Croubels S. In vitro adsorption and in vivo pharmacokinetic interaction between doxycycline and frequently used mycotoxin binders in broiler chickens. J Agric Food Chem. 2015;63:4370–4375. doi:10.1021/acs.jafc.5b00832.

[86] Rosa CA, Miazzo R, Magnoli C, Salvano M, Chiacchiera SM, Ferrero S, Saenz M, Carvalho EC, Dalcero A. Evaluation of the efficacy of bentonite from the south of Argentina to ameliorate the toxic effects of AF in broilers. Poult Sci. 2001;80:139–144. doi:10.1093/ps/80.2.139.

[87] Vekiru E, Fruhauf S, Sahin M, Ottner F, Schatzmayr G, Krska R. Investigation of various adsorbents for their ability to bind Aflatoxin B1. Mycotoxin Res. 2007;23:27–33. doi:10.1007/BF02946021.

[88] Ciegler A, Lillehoj B, Peterson E, Hall HH. Microbial detoxification of aflatoxin. Appl Microbiol. 1966;14:934–939.

[89] Mc Cormick SP. Microbial detoxification of mycotoxins. J Chem Ecol. 2013;39:907–918.

[90] Molnar O, Schatzmayr G, Fuchs E, Prillinger H. Trichosporon mycotoxinivorans sp. nov., a new yeast species useful in biological detoxification of various mycotoxins. Syst Appl Microbiol. 2004;27:661–671. doi:10.1078/0723202042369947.

[91] Schatzmayr G, Zehner F, Taubel M, Schatzmayr D, Klimitsch A, Loibner AP, Binder EM. Microbiologicals for deactivating mycotoxins. Mol Nutr Food Res. 2006;50:543–551. doi:10.1002/mnfr.200500181.

[92] Hartinger D, Moll WD. Fumonisin elimination and prospects for detoxification by enzymatic transformation. World Mycotoxin J. 2011;4:271–283. doi:10.3920/WMJ2011.1285.

[93] Grenier B, Applegate TJ. Modulation of intestinal functions upon mycotoxin ingestion: meta-analysis of published experiments in animals. Toxins. 2013;5:396–430. doi:10.3390/toxins5020396.

[94] Prakash B, Kedia A, Kumar Mishra P, Dubey NK. Plant essential oils as food preservatives to control moulds, mycotoxin contamination and oxidative deterioration of agri-food commodities – potentials and challenges. Food Control.2015;47:381–391. doi:10.1016/j.foodcont.2014.07.023.

[95] Abdel-Wahhab MA, Hassan NS, El-Kady AA, Khadrawy YA, El-Nekeety AA, Mohamed SR, Sharaf HA, Mannaa FA. Red ginseng extract protects against aflatoxin B1 and fumonisins-induced hepatic pre-cancerous lesions in rats. Food Chem Toxicol. 2010;48:733–742. doi:10.1016/j.fct.2009.12.006.

[96] Manafi M, Khosravinia H. Effects of aflatoxin on the performance of broiler breeders and its alleviation through herbal mycotoxin binder. J Agr Sci Tech. 2013;15:55–63.

Identification of Microbial and Gaseous Contaminants in Poultry Farms and Developing Methods for Contamination Prevention at the Source

Dorota Witkowska and Janina Sowińska

Abstract

Microbial concentrations in poultry houses increase over time and contribute to the sick building syndrome. Very high and often logarithmic growth rates are reported for aerobic mesophilic bacteria, which account for the majority of known pathogenic bacteria. Bioaerosols suspended in air also contain mold spores and mold fragments, mostly fungi of various genera, including pathogenic fungi that produce mycotoxins. Microbiological mineralization of organic compounds, processes that involve litter and fecal microbes, produces toxic gases, including ammonia, carbon dioxide (CO_2), as well as volatile toxic and aroma compounds. The above threats have led to the initiation of various measures to limit pollution at the source, including legal regulations and methods aiming to neutralize the adverse effects of pollution (dietary, production, and hygiene standards). Hygienic methods are recommended as alternative methods of reducing contamination in poultry houses. Essential oil mist, organic and organic-mineral biofilters, litter additives, such as aluminosilicates (bentonite, vermiculite, halloysite), microbiological and disinfecting preparations, herbal extracts, and calcium compounds may improve hygiene standards in poultry farms.

Keywords: bacteria, fungi, gases, poultry farm, reduction methods

1. Introduction

The specific microclimate of farm buildings promotes the accumulation of bioaerosols containing harmful biological substances. Microbial concentrations in poultry houses increase over time and contribute to the sick building syndrome. Very high and often logarithmic growth rates are reported for aerobic mesophilic bacteria, which account for the majority of known pathogenic bacteria. In addition to Gram-positive cocci (*Staphylococcus, Enterococcus*) and bacilli (*Bacillus*), other aerobic mesophilic bacteria include Gram-negative bacteria of the family *Enterobacteriaceae*, including *Escherichia coli*, *Salmonella* sp., *Shigella* sp., *Enterobacter* sp., *Proteus* sp., and *Klebsiella* sp., as well as *Pseudomonas* sp., *Acinetobacter* sp., *Flavobacterium* sp. [1–4]. The cell membranes of Gram-negative bacteria contain pro-inflammatory and allergenic lipopolysaccharide complexes, known as endoxins, which are released after the death of the bacterial cell. High endoxin concentrations are also observed in poultry houses [5]. Bioaerosols suspended in air also contain mold spores and mold fragments, mostly fungi of various genera (*Penicillium, Aspergillus, Fusarium, Mucor, Trichosporon, Alternaria, Cladosporium, Trichophyton, Epicoccum*, etc.), including pathogenic fungi that produce mycotoxins [1, 4, 6]. Pathogenic yeasts (*Candida albicans* and *Cryptococcus neoformans*) are also frequently identified in animal houses. Enzymatic and microbiological mineralization of organic compounds, processes that involve litter and fecal microbes, produces toxic gases, including ammonia and carbon dioxide (CO_2), as well as volatile toxic and aroma compounds such as cyclic hydrocarbons, aldehydes, ketones, alcohols, free fatty acids, mercaptans, esters, phenols, cyclic amines, and sulfides [7–8].

The combined effect of airborne pollutants is one of the key stressors in poultry farms. High concentrations of microorganisms, endotoxins, mycotoxins, gas, and dust exert adverse effects on the structure and protective functions of mucous membranes, in particular in the respiratory system and the conjunctiva, leading to allergic reactions, inflammations, and increasing susceptibility to infectious diseases. Numerous studies have demonstrated that excessive ammonia concentrations in hen house lower productivity [2, 9–11]. Airborne pollutants also have an adverse influence on farm employees [2]. The pollutants emitted by poultry farms have negative environmental consequences. Nitrogen compounds contaminate soil and ground waters [12], whereas gases such as CO_2, CH_4, and N_2O contribute to the greenhouse effect [13, 14]. Recent years have witnessed an increasing interest in volatile organic compounds (VOCs), which are odor-producing compounds [15, 16].

The above threats have led to the initiation of various measures to limit pollution at the source, including legal regulations (international conventions and the resulting legal acts that are binding for the signatory countries) and methods aiming to neutralize the adverse effects of pollution (dietary, production, and hygiene standards). The concentrations of biological and chemical air pollutants vary significantly between farms and livestock facilities, and they are determined not only by the animal species, but also by the housing and management system. For this reason, safe pollution thresholds are very difficult to define. Guidelines for limiting the exposure to selected chemical and physical factors have been developed in occupational medicine, but general threshold limit values (TLV) for biological compounds are very difficult to establish due to an absence of epidemiological data describing the correlations between exposure and health consequences [17]. Different organisms have varied susceptibility to toxic substances; there is a general absence of standardized measurement methods and a scarcity of source data relating to the most widespread bioaerosols, which further exacerbates the problem. Threshold values for farm buildings are even more difficult to determine due to a higher number of limiting factors. Air pollution poses a serious health threat for animals and farm employees; therefore, new research into the type and concentrations of airborne pollutants in various housing systems is needed to effectively mitigate the problem.

This manuscript reviews the results of our previous work and other studies into quantitative and qualitative identification of microbial and gaseous contaminants in poultry houses and methods for the prevention of contamination at the source.

2. Microbial contaminations of poultry farms

The level of microbial contamination in poultry houses is one of the most important sanitary and hygienic indicators. The main sources of microorganisms in poultry houses are birds, their excrements, feed, litter, ventilation air, and even employees. Microbes carried by dust, water vapor, and secretions from the respiratory tract form bioaerosol. Birds breathe air which acts as a major vector for microorganisms. Most microbes are saprophytes, but some airborne microorganisms may be pathogenic. Pathogens that enter the respiratory system with liquid droplets and dust may cause infections. The smallest particles measuring <50 nm pose the greatest epizootic risk because they are slowly deposited and spread even at low air flow rates. The flock is constantly exposed to pathogenic bioaerosols when sick or infected birds are present in the poultry house [18].

Microbial survival is determined by temperature, humidity, and other environmental parameters. Relative humidity in poultry houses generally does not support bacterial proliferation (the 50–80% range is lethal for bacteria), and microbial contamination of air, litter, and surfaces in poultry farm buildings can be attributed mainly to high flock density and the continued presence of microbial sources. Poultry farms are significant pollutants of the external environment, and they could pose an epidemiological risk if biosecurity principles are not observed.

The microbial concentrations reported inside and outside poultry farms (**Tables 1** and **2**) differ considerably in the literature [4, 19, 20–26]. Our previous work and other studies revealed aerial contamination in the range of 3.1–6.4 $\log_{10}$ cfu/m^3 in broiler houses, 4.5–7.6 $\log_{10}$ cfu/m^3 in turkey houses, and 4.7–8.3 $\log_{10}$ cfu/m^3 in laying hen houses. Fungal concentrations in broiler, hen, and turkey houses were determined at 4.0–5.9, 3.8–5.8, and 2.7–5.5 $\log_{10}$ cfu/m^3, respectively. Outdoor concentrations were reported at 0–5.6 $\log_{10}$ cfu/m^3 for bacteria and 0–4.8 $\log_{10}$ cfu/m^3 for fungi, depending on the distance. Microbial contamination levels are influenced by various factors, including bird species, stocking density, season, ventilation system, microclimate, and litter quality.

Witkowska et al. [19] studied the total counts of aerobic mesophilic bacteria and fungi in fresh litter and in the air in a broiler house under changing temperature and humidity conditions, and changing physicochemical properties of litter throughout the rearing period. The total counts of aerobic mesophilic bacteria and fungi in fresh litter tended to increase during the rearing period, to reach 9 and 8 $\log_{10}$ cfu/g, respectively, in the last week. An insignificant increase in litter pH was also noted throughout the experiment, which—combined with increasing excreta amounts and fermentation processes in fresh litter—could promote microbial growth. The above factors enhanced ammonia production in litter (6 mg/kg at the beginning of the experiment vs. 12 mg/kg in the last week). Despite a gradual decrease in indoor temperature accompanied by an increase in humidity, microbial air contamination did not follow the same pattern as litter contamination. Bacterial and fungal counts varied between weeks of the rearing period, most likely due to changes in dust levels and ventilation efficiency. Bacterial counts were lowest in week 3 (4.6 $\log_{10}$ cfu/m^3) and highest at the end of rearing (5.3 $\log_{10}$ cfu/m^3). Fungal counts were lowest at the beginning of the experiment (4.2 $\log_{10}$ cfu/m^3) and highest in weeks 2 and 5 (4.7 $\log_{10}$ cfu/m^3). Lawniczek-Walczyk et al. [27] observed a significant increase in the concentrations of bacterial and fungal aerosols and endotoxins in chicken houses in successive stages of production. They also reported seasonal correlations in the size of bacterial populations. The concentrations of airborne bacteria were significantly higher in summer than in winter.

Flock (no. of birds, rearing period)	Housing type (no. of buildings, ventilation system, type of bedding, season)	Total microorganisms level ($\log_{10}$ cfu/m^3) mean range (min-max)		References
A. Bacteria				
Broilers (230 400 birds, 8 weeks)	12 buildings, mechanical ventilation, sawdust or straw litter	4.8 (3.1–5.2)		Baykov and Stoyanov [20]
Broilers (5300 birds, six weeks)	One building, mechanical ventilation, sawdust or wood shaving litter, spring	5.1 (4.2–5.3)		Vučemilo et al. [4]
Broilers (350 birds, five weeks)	One building, mechanical ventilation, straw litter, winter	5.1 (5.1–5.3)		Witkowska et al. [19]
Broilers (360 birds, six weeks)	One building, mechanical ventilation, straw litter, summer and winter	Summer 5.9 (5.0–6.2)	Winter 6.0 (4.9–6.4)	Wójcik et al. [21]
Broilers (41 000 birds, six weeks)	Two buildings, mechanical ventilation, straw litter, spring/summer	– (5.1–5.7)		Lonc and Plewa [22]
Lying hens (19 500 birds)	Three buildings mechanical ventilation, straw litter, spring–autumn	7.8 (4.7–8.3)		Bródka et al. [23]
Turkeys (2000 birds)	One building, Louisiana-type, wood chips or straw litter	6.9 (4.5–7.6)		Saleh et al. [26]
B. Fungi				
Broilers (5300 birds, 6 weeks)	One building, mechanical ventilation, sawdust or wood shaving litter, spring	4.5 (4.0–4.9)		Vučemilo et al. [4]

Flock (no. of birds, rearing period)	Housing type (no. of buildings, ventilation system, type of bedding, season)	Total microorganisms level ($\log_{10}$ cfu/m^3) mean range (min-max)		References
Broilers (350 birds, 5 weeks)	One building, mechanical ventilation, straw litter, winter	4.5 (4.2–4.7)		Witkowska et al. [19]
Broilers (360 birds, 6 weeks)	One building, mechanical ventilation, straw litter, summer and winter	Summer 5.3 (4.6–5.8)	Winter 5.5 (4.7–5.9)	Wójcik et al. [21]
Broilers (41 000 birds, 6 weeks)	Two buildings, mechanical ventilation, straw litter, spring/summer	– (4.6–5.0)		Lonc and Plewa [22]
Lying hens (19 500 birds, 1 year)	Three buildings, mechanical ventilation, straw litter, spring–autumn	5.3 (3.8 Spr–5.8 Aut)		Sowiak et al. [24]
Turkeys (2000 birds)	One building, Louisiana-type, wood chips or straw litter	5.0 (2.7–5.5)		Saleh et al. [26]

Table 1. Bioaerosol concentrations in poultry houses.

Type of farm (no. of birds)	Total microorganisms level ($\log_{10}$ cfu/m^3) mean range (min–max)	Distance	References
A. Bacteria			
Broilers (19 200)	– (2.3–5.6)	3 km–10 m	Baykov and Stoyanov [20]
Broilers (350)	2.6 (0–2.9)	3 m	Witkowska et al. [19]
Broilers (360)	3.9 (0–4.4)	3 m	Wójcik et al. [21]
Broilers (41 000)	– (1.6–3.9)	–	Lonc and Plewa [22]
Broilers (23 000)	– (3.7–4.1)	125–10 m	Plewa-Tutaj et al. [25]
B. Fungi			
Broilers (350)	3.0 (0–3.2)	3 m	Witkowska et al. [19]
Broilers (360)	3.8 (0–4.3)	3 m	Wójcik et al. [21]
Broilers (41 000)	– (1.3–4.1)	–	Lonc and Plewa [22]
Lying hens (19 500)	4.6 (4.2–4.8)	–	Sowiak et al. [24]

Table 2. Bioaerosol concentrations around poultry houses.

Bacterial and fungal species and serotypes isolated from poultry farms are presented in **Table 3**. Numerous studies [4, 19–25] revealed that bioaerosols from poultry houses contain Gram-positive bacteria, including *Streptococcus, Staphylococcus, Micrococcus, Enterococcus, Aerococcus, Corynebacterium, Brevibacterium, Cellulomonas* and *Bacillus*, as well as Gram-negative bacteria, including *Escherichia, Enterobacter, Klebsiella, Proteus, Citrobacter, Pasteurella, Pantoea,*

Moraxella, and *Pseudomonas*. Witkowska et al. [19] report that in their investigation, molds (*Fusarium, Penicillium, Aspergillus*, and many others) predominated in the air and litter at the beginning of the production cycle, whereas yeast counts increased to 90–100% towards the end of the experiment, particularly in litter.

Species	Serotypes	Inside houses	Outside houses	References
A. Bacteria				
Streptococcus	*S. pyogenes*	+	−	Baykov and Stoyanov 1999 [20]; Vučemilo et al. [4]; Lonc and Plewa 2011 [22]; Bródka et al. 2012 [23]; Lawniczek-Walczyk et al. 2013 [27]; Plewa-Tutaj et al. 2014 [25]
	S. bovis	+	−	
	S. mitis	+	+	
Staphylococcus	*S. xylosus*	+	+	
	S. hyicus	+	−	
	S. saprophyticus	+	−	
	*S. aureus**	+	+	
	S. epidermidis	−	+	
	S. lentus	+	+	
	S. sciuri	+	+	
	S. chromogenes	+	−	
	S. cohnii	−	+	
Micrococcus	*M. sedentarius*	+	−	
	M. luteus	+	+	
	M. lylae	−	+	
	M. halobius	−	+	
Enterococcus	*E. faecalis*	+	−	
	E. faecium	+	−	
	sp.	−	+	
Aerococcus	*A. viridans*	+	−	
Corynebacterium	*C. xerosis*	+	+	
Brevibacterium	*sp.*	+	−	
Cellulomonas	*C. cellulans*	+	−	
Bacillus	*sp.*	+	−	
	mycoides	−	+	
Escherichia	*E. coli*	+	+	
Enterobacter	*E. sakazakii*	+	−	
	E. agglomerans	+	+	
	*E. cloacae**	+	+	
Klebsiella	*K. pneumoniae**	+	−	
Shigella	*S. boydii*	−	+	
Proteus	*P. mirabilis**	+	+	

Species	Serotypes	Inside houses	Outside houses	References
Citrobacter	C. farmerii	+	+	
Pasteurella	sp.	+	−	
Pantoea	sp.	+	−	
Moraxella	sp.	+	−	
Providencia	sp.	−	+	
Pseudomonas	P. aeruginosa	+	+	
	P. fluorescens	+	+	
	P. alcaligenes	+	−	
	P. stutzeri	+	−	
	P. chlororaphis	−	+	
Xantomonas	X. maltophila	−	+	
B. Fungi				
Penicillium	P. notatum	+	−	Vučemilo et al. 2007 [4]; Witkowska et. al. 2010 [19]; Wójcik et al. 2010 [21]; Lonc, Plewa 2011 [22]; Sowiak et al. 2012 [24]; Lawniczek–Walczyk et al. 2013 [27]
	P. expansum	+	−	
	P. olivinoviridae	+	−	
	P. claviforme	+	−	
	P. viridicatum	+	−	
	P. chrysogenum	−	+	
Aspergillus	A. niger	+	+	
	A. nidulans	+	+	
	A. ochraceus	+	−	
	A. oryzae	+	−	
	8A. candidus	+	−	
	A. fumigatus*	+	−	
	A. glaucus	−	+	
	A. parasiticus	−	+	
	A. clavatus	−	+	
Fusarium	F. oxysporum	+	+	
	F. graminearum	+	+	
Geotrichum	sp.	+		
Scopulariopsis	S. brevicaulis	+	+	
	S. acremonium	+	+	
Alternaria	A. alternata	+	+	
	A. tennuissima	+	+	
Trichoderma	T. viridae	+	+	
Drechslera	D. graminae	+	+	
Mucor	M. mucedo	+	+	
Rhizous	R. oryzae	+		

Species	Serotypes	Inside houses	Outside houses	References
	R. stolonifer	+		
	R. nodosus	+		
Cladosporium	C. cladosporoides	+	+	
Candida	C. albicans	+	−	
	C. inconspicua	+	−	
	C. lambica	+	+	
	C. famata	+	−	
	C. pelliculosa	+	−	
Cryptococcus	C. laurentii	+	−	
	C. humicola	+	−	
	sp.	−	+	
Acremonium	A. strictum	+		
Trichophyton	T. mentagrophytes*	+		
Ulocladium	sp.	−	+	
Verticilium	sp.	−	+	
Scedosporium	sp.	−	+	
Mycelia	M. sterilia	−	+	
Rhodotorula	R. rubrum	−	+	

*Microorganisms classified into group 2 according to level of risk of infection [27].

Table 3. The most common microorganisms isolated from poultry farms.

Some microbial species and serotypes, such as *Staphylococcus aureus*, *Enterobacter cloacae*, *Klebsiella pneumoniae*, *Proteus mirabilis*, *Trichophyton mentagrophytes*, and *Aspergillus fumigatus*, are pathogenic for animals and humans. Many bacteria and fungi are opportunists which are particularly dangerous for organisms with compromised immunity. Low hygiene standards and high levels of microbial and gaseous contamination may synergistically contribute to lower immunity and susceptibility to infections. The presence of microorganisms such as *Brevibacterium*, *Alternaria*, and *Cladosporium* in poultry houses indicates that microbes from the external environment, including soil, can spread to farm buildings. The prevalence of pathogenic microorganisms outside poultry buildings, even several kilometers away from the site, also indicates that ventilation air may contaminate the external environment.

Broiler houses are particularly infested by fungi of the genera *Penicillium*, *Aspergillus*, and *Fusarium*, which are the main fungi producing pathogenic mycotoxins such as T-2 toxin, aflatoxin, ochratoxin, and zearalenone. Other toxigenic fungi species, for example, *Alternaria*, *Cladosporium*, *Trichoderma*, *Rhizopus*, *Stachybotrys*, have also been identified in poultry buildings. Even low concentration of mycotoxins is known to cause immunosuppression, allergies, inflammation of the respiratory tract, and they may have impact on growth parameters of birds [4, 6, 19, 22, 23, 25, 27–30].

Bioaerosols also contain suspended endotoxins. These lipopolysaccharide complexes are associated with the cell membrane of Gram-negative bacteria which are released after the death of bacterial cells. High endoxin concentrations are also observed in poultry houses. Inhaled bioaerosol particles carrying endotoxins have pro-inflammatory and allergenic properties and may lead to chronic respiratory diseases in poultry [5]. The mean concentrations of endotoxins in aerosol fractions from poultry houses were determined in the range of 11.2–3406 ng/m^3 and were significantly higher than in other livestock buildings [3, 5, 27]. Seedorf et al. [31] observed the highest endotoxin concentrations in laying hen houses. Lawniczek-Walczyk et al. [27] observed that endotoxin concentrations in poultry houses increased significantly in successive stages of production. The above authors concluded that high levels of airborne microorganisms and their bioproducts could pose a serious risk of respiratory diseases. For this reason, widely accepted guidelines for hygiene evaluation need to be established in poultry farms.

In Poland, the proposed threshold limit values (TLV) are 5.0 $\log_{10}$ cfu/m^3 for bacteria and 4.7 $\log_{10}$ cfu/m^3 for fungi [27], but those limits apply to bioaerosol concentrations in employee facilities. Krzysztofik [32] recommended a limit of 5.0 $\log_{10}$ cfu/m^3 for bacteria and a more restrictive threshold value for fungi at 3.3 $\log_{10}$ cfu/m^3. The TLV for endotoxins recommended by the Polish Expert Committee for Biohazards in Indoor Environments is 200 ng/m^3 [18, 27]. There are no guidelines for poultry houses.

3. Gaseous contaminations of poultry farms

The composition of air in poultry houses significantly differs from atmospheric air. In addition to basic gaseous components (N_2—nitrogen, O_2—oxygen, Ar—argon, and CO_2—carbon dioxide), the air inside poultry houses also contains compounds that are not normally found in atmospheric air. Birds, their excrements, feed, and process equipment are the main sources of volatile chemical compounds in poultry houses. Ammonia and carbon dioxide are most frequently encountered in farm buildings, and they contribute to the risk of disease if present in excessive concentrations. For this reason, ammonia and carbon dioxide are regarded as the most toxic gases in poultry houses.

Carbon dioxide (CO_2) is a natural component of air, and its concentrations generally do not exceed 300 ppm (0.03%). Carbon dioxide is responsible for breathing control in the respiratory system. In densely stocked poultry houses, carbon dioxide concentrations are significantly higher than in atmospheric air, but they should not exceed 2000 ppm. At higher concentrations in poultry houses, CO_2 weakens respiratory defense mechanisms and increases susceptibility to respiratory diseases. Carbon dioxide poses a serious hazard to health and life at concentrations higher than 10 000 ppm. Carbon dioxide levels in poultry houses are a robust indicator of ventilation efficiency. Its concentrations increase rapidly in poorly ventilated buildings.

In poultry houses, ammonia (NH_3) is released from excreta which contain nitrogen in the form of uric acid. Ammonia is produced in the process of microbial fermentation. Ammonia production increases in conditions that support microbial proliferation, including high temperature, high humidity, high pH, and presence of organic matter. In poultry houses, ammonia concentrations should not exceed 13 ppm for adult birds and 10 ppm for chicks. At higher concentrations, NH_3 can compromise growth, whereas exposure to more than 30 ppm of ammonia can lead to respiratory dysfunctions such as intensified mucus secretion, shallow breathing, and bronchoconstriction. High levels of ammonia can impair immunity and increase susceptibility to respiratory infections and ocular abnormalities in poultry.

Ammonia and other nitrogen compounds (NO_x) originating from poultry production contaminate soil and groundwater. Some gaseous compounds emitted by poultry farms, in particular CO_2 and NO_x, are greenhouse gases which contribute to global warming. Many volatile compounds are classified as odors, and hundreds of volatile organic compounds (VOCs) are identified in poultry houses. There are no guidelines concerning the odor detection threshold or VOCs' impact on odor formation in poultry farms due to the scarcity of simple instruments for measuring air contamination in the production process. Research studies revealed the presence of aromatic and aliphatic hydrocarbons, aldehydes, ketones, alcohols, free fatty acids, mercaptans, esters, phenols, cyclic amines, nitriles, and sulfur compounds in bird farms [7–8, 13, 15, 18, 33].

Witkowska [33] conducted qualitative and quantitative identification of gaseous contaminants on a commercial turkey farm by Fourier transform infrared spectroscopy (FTIR). It was found that ammonia and carbon dioxide were the predominant gases in turkey houses, and both were present throughout the entire growth cycle, which is consistent with the findings of other authors. The mean concentrations of carbon dioxide and ammonia were 220–2058 ppm (min = 176 ppm, max = 2460 ppm) and 4–31 ppm (min = 4 ppm, max = 58 ppm), respectively. The highest carbon dioxide concentrations (approx. 2000 ppm) were reported in farm buildings in weeks 4 and 7, and a significant decrease in 600 ppm was observed in week 10. A lessening tendency was noted until the end of the production cycle; in week 19, mean carbon dioxide concentrations in turkey houses were approximately 90% lower than at the beginning of the experiment. Average ammonia concentrations increased from 7 ppm at the beginning of the study to over 30 ppm in week 7. A significant decrease in ammonia levels was observed in subsequent weeks. The decreasing trend was sustained until the end of the rearing period, and the mean concentrations of NH_3 were 90% lower in the second half of the cycle. The decrease resulted from higher ventilation and air exchange rates in turkey houses (at the last stage of the study, and the rate of ventilation was 10-fold higher than at the initial stage). The increase in CO_2 and NH_3 levels in week 7 was related to diet modification and increased excreta moisture. Thiols, nitriles, amines, aldehydes, hydrocarbons, and other volatile organic and inorganic compounds were also identified in the air inside the buildings, but they were emitted periodically and their mean concentrations were significantly lower in comparison with CO_2 and NH_3. In contrast to the majority of other contaminants, nitrogen compounds (nitriles, amines, aldehydes) and some hydrocarbons (chloroethane, 1.3-butadiene) were present at higher concentrations in the second half of the production cycle. During the experiment, trace amounts of alcohols, organic acids, ketones, phenols, nitrogen oxides, and sulfur oxides were also detected in the air inside farm buildings. Mixtures of those compounds act as odorants even at low concentrations.

According to EU directives and Polish regulations, ammonia concentrations in broiler and laying hen houses should not exceed 20 ppm, and carbon dioxide concentrations should be limited to 3000 ppm. Gas concentrations have to be kept within safe limits in turkey, duck, and geese houses [33]. More restrictive limits have been recommended by some authors [18], and further research is needed to determine tolerable limits for different poultry species and rearing systems.

4. Hygienic methods for reducing contamination in poultry houses

The search for effective, inexpensive, and environmentally friendly methods of lowering contamination levels in poultry production has continued for many years. Ventilation systems play a key role in maintaining the optimal microclimate in poultry houses, and devices that generate negative ions remove dust and moisturize air can also be installed in farms to limit air pollution. Unfortunately, such solutions are relatively expensive, and they are not widely used in poultry production.

Witkowska and Sowińska [34] study aimed to assess the antibacterial effects of natural essential oils (peppermint oil—PO and thyme oil—TO) in broiler houses. The results of the study demonstrated that essential oil mist may improve hygiene standards in broiler farms. The mean total counts of aerobic mesophilic bacteria in the control room were significantly higher than in rooms treated with essential oils—5.8 $\log_{10}$ cfu/m^3 vs. 5.6 $\log_{10}$ cfu/m^3 (PO) and 5.5 $\log_{10}$ cfu/m^3 (TO). A similar trend was observed with regard to wall contamination—total mesophilic counts ranged from around 2.4 $\log_{10}$/100 cm^3 in rooms fogged with essential oils to 3.3 $\log_{10}$/100 cm^3 in the control room, and the statistic differences between control and experimental groups were determined. Total bacterial counts on drinker surfaces in rooms fogged with essential oils were lower than in the control room (PO–4.6 $\log_{10}$, TO–4.3 $\log_{10}$). Average drinker contamination was significantly (by 0.5 $\log_{10}$) higher in the control room than in the room fogged with thyme oil. The average total count of litter bacteria ranged from 8.9 $\log_{10}$ cfu/g in the control group to 8.2 $\log_{10}$ cfu/g in the thyme oil group, and the noted difference was statistically significant. Litter contamination was also lower in the room fogged with peppermint oil (8.5 $\log_{10}$ cfu/g), compared with the control room, but the difference was not significant. An analysis of extreme values of bacterial counts in the air on the walls and drinkers and in litter revealed that bacterial contamination levels were effectively reduced by essential oils. The average counts of bacteria of the family *Enterobacteriaceae* and mannitol-positive *Staphylococcus* in the air on wall and drinker surfaces were lower in experimental rooms than in the control room. A similar tendency was noted with respect to the counts of *Staphylococci* in litter, but no significant differences were found between groups. The counts of coliforms were lowest in the room fogged with thyme oil, and they were higher in the room treated with peppermint oil than in the control room. Both oils reduced bacterial counts, but thyme oil was more effective in eradicating *Enterobacteriaceae*, whereas peppermint oil had a higher inhibitory effect on the proliferation of *Staphylococci*.

According to Tymczyna et al. [3, 7, 35, 36], biofilters offer a relatively cheap and effective solution for poultry farms. Biofilters are containers with many partitions that house a high-pressure fan, an air moisturizing chamber and a biofiltration chamber. The biofiltration chamber is a bed of various media, such as peat, compost, horse manure, and wheat straw. Toxic gases are partially or completely biodegraded by bacteria that occur naturally in bed media or are artificially introduced to substrates. Bacterial proliferation is influenced by the parameters of bed media, including fertility, moisture content, temperature, and pH. The cited authors demonstrated that ammonia and other toxic compounds (nitrates, nitrites, sulfates, chlorides, phosphates) present in ventilation systems can be effectively eliminated with the use of open biofilters in laying hen farms. Biological beds composed of peat, treated compost, horse manure, and wheat straw reduced ammonia concentrations by 36–89% (68.6% on average) and eliminated other harmful compounds in 66–100%.

Chmielowiec-Korzeniowska et al. [37] evaluated the effectiveness of a prototype container biofilter in eliminating organic air pollutants in a chick hatchery. The biofilter bed was composed of sallow peat (30%), fibrous peat (30%), treated compost (10%), fermented horse manure (10%), and wheat straw (20%). The tested device decreased the levels of all pollutants by 66% on average, and it was most effective in removing hexanal (95%) and toluene (76%).

The same team of researchers [3] evaluated the effectiveness of organic and organic-mineral biofilters in eliminating Gram-negative bacteria, dust, and bacterial endotoxins from exhaust air leaving a chick hatchery. All evaluated filters were effective in removing bacterial aerosols and somewhat less effective in reducing dust pollution. Endotoxins were not effectively eliminated. A biofilter with an organic-mineral bed containing 20% halloysite, 40% compost, and 40% peat was most effective in lowering contamination levels.

Numerous research studies demonstrated that aluminum silicates can be effectively used as bed media in air filtering devices. An important advantage of aluminum silicates is that they are relatively cheaper and less toxic for animals and the environment than commercially available chemical sorbents.

Opaliński et al. [38] evaluated the ability of selected aluminum silicates to absorb ammonia. The tested substrates were raw halloysite, roasted halloysite, activated halloysite, raw bentonite clay, and expanded vermiculite (EV). The experiment was conducted under strictly controlled laboratory conditions. The analyzed substrates' sorptive capacity was determined based on differences in ammonia concentrations in a stream of air before and after it passed a sorptive bed with a known volume. All evaluated sorbents lowered ammonia concentrations in air. The most effective sorbent was activated halloysite, followed by raw halloysite, roasted halloysite, and raw bentonite, whereas vermiculite was least effective in capturing ammonia.

Opaliński et al. [39] also analyzed the ability of selected aluminum silicates to eliminate noxious odors in conditions similar to those found in animal facilities. Chicken droppings were placed in fertilizer chambers, and the odor capturing abilities of raw halloysite, roasted halloysite activated halloysite, raw bentonite, roasted bentonite, and expanded vermiculite were evaluated after 24 h. Ammonia was most effectively removed (81%) by activated halloysite, whereas roasted halloysite was the least effective sorbent. In addition to ammonia, the analyzed air samples also contained 24 odorous volatile compounds, including five toxic substances. All of the tested aluminum silicates effectively decreased the concentrations of the identified compounds, and their average sorptive capacity ranged from 56% for raw halloysite to 84% for roasted bentonite. Roasted bentonite reduced the levels of seven odorous compounds by more than 90% and eliminated IH-indole, dimethyl trisulfide, and pyridine in even 100%.

Organic and mineral compounds are added to litter to improve its quality [40, 41]. The objective of Korczyński et al. [42] study was to determine the effectiveness of expanded vermiculite (EV) and raw halloysite (HS) in reducing the emissions of ammonia and volatile organic compounds (VOCs) from litter in turkey houses. Mean ammonia concentrations were lower in the sectors where the analyzed sorbents were used. The average differences in NH_3 levels between the control sector and the sectors where vermiculite and halloysite were added to litter reached 15.1 and 14.6%, respectively. The highest efficacy of both sorbents was noted in the first week of the study (statistically significant differences). The application of halloysite and vermiculite decreased ammonia concentrations by 38 and 25%, respectively, compared with the control sector. Similar trends were observed in the subsequent 2 weeks, but differences in ammonia concentrations between the control sector and experimental sectors were much lower (3.4–11.4%) and statistically non-significant. A total of 15, 14, and 11 volatile organic compounds were identified in the air in the control, HS and EV sectors, respectively. Pentadecane, 1-phenylethanone, dimethyl tetrasulfide, and 4-hydroxytoluene were detected in the control sector, but they were not found in experimental sectors. Methylbenzene and 4-methyl-2-heptanone were identified in the air in the sectors where the sorbents were used, and 2-undecanone was detected in the HS sector—those compounds were not found in the control sector. Chlorobenzene was the predominant VOC in the air in all sectors. Sorbents added to litter were most effective in reducing the emissions of compounds with more complex molecular structure. VOC levels decreased by 73.4 and 83.1% following the use of halloysite and vermiculite, respectively.

Manafi et al. [43, 44] observed that high grade sodium bentonite in diet reduced the toxicity of aflatoxin and marginally ameliorated the effect of ochratoxin A and aflatoxin B_1 in broilers.

In a search for effective methods to reduce contamination levels in poultry production, various litter additives were analyzed [28, 45–47], including a microbiological preparation (Biosan-GS®) and disinfecting preparations (Lubisan®, Stalosan F®, Profistreu®). All additives contributed to a decrease in litter pH and moisture content [28, 46] thus reducing ammonia concentrations in the air and litter [45, 46], and microbial air contamination levels in poultry houses [28]. Birds kept on "optimized" litter were characterized by higher body weight gains [46] and lower culling and mortality rates, at similar feed intake levels. An analysis of internal organs (liver, spleen, kidneys, lungs, cornea) and selected blood parameters showed that the above litter additives were safe and posed no threat to bird health [45, 47].

Saponin extracts from the South American plants of Mojave yucca (*Yucca schidigera*) and soap bark tree (*Quillaja saponaria*) are added to feed and litter in poultry farms. Saponins block bacterial urease and slow down urea decomposition in the uricolytic cycle. They increase the availability of feed protein for birds and decrease the excretion of nitrogen compounds that can be converted to ammonia [48, 49].

Litter can also be disinfected with calcium compounds before animals are introduced to a farm building [50]. Many authors demonstrated that the addition of calcium oxide to poultry litter significantly decreases bacterial counts, in particular *Salmonella* which continues to pose a serious problem for the producers and consumers of poultry meat and eggs [51, 52].

In the literature, there are no recommendations regarding the optimal doses of calcium additives in litter, but calcium compounds are popularly used in poultry farms on account of their low cost. Despite the above, calcium additives should be applied with caution because exposure to excessive calcium concentrations can irritate or burn mucosal membranes and the skin. Calcium oxide reacts with water to increase temperature, which can harm birds reared on litter. Mituniewicz [53] attempted to determine the optimal doses of calcium compounds (CaO and CaOMgO) which can be safely added to litter before birds are introduced to poultry houses. The cited author analyzed the physicochemical and microbiological parameters of litter, microclimate conditions, selected blood biochemical parameters and bird performance to conclude that a single application of 250 g CaO or CaOMgO per square meter of litter delivered the best results. Calcium compounds had a positive effect on the physicochemical parameters of litter and microbial counts. The tested additives, in particular calcium oxide, led to a significant increase in litter temperature within the safe limits. Calcium oxide also lowered the relative moisture content of litter, in particular in the last weeks of the rearing period. The combination of calcium oxide and magnesium oxide induced a greater improvement in the analyzed parameters. Calcium compounds were effective disinfectants which reduced the counts of incubated yeasts already in the third week of the experiment. Ammonia concentrations were significantly lowered in a poultry house where calcium compounds were added to litter. The results of blood serum biochemistry analyses revealed that calcium compounds did not exert a negative effect on the birds' health. Chickens reared on litter with calcium additives were characterized by higher weight gains and improved performance.

Calcium peroxide (CaO_2) is an inorganic compound and a source of oxygen. This compound is sparingly soluble in water, and hydrogen peroxide, a source of free radicals (chemical oxidation) and oxygen, is gradually released during the slow decomposition of CaO_2. The above creates a supportive environment for aerobic microorganisms [54]. Piotrowska [55] attempted to determine the optimal dose at which calcium oxide should be combined with litter to improve hygiene conditions in broiler houses and broiler performance. During a four-week laboratory experiment (without birds) involving analyses of the qualitative parameters of chicken litter and microclimate conditions in broiler houses, the cited author determined the optimal dose of CaO_2 at 2 g m^{-2} litter. The above dose was then tested under production conditions in a poultry farm. The surface temperature of the experimental litter was reduced by 1°C, and its moisture content decreased in comparison with the control litter (57.11% vs. 70.33%), which lowered the counts of aerobic mesophilic bacteria. Ammonia concentrations in the experimental poultry house did not exceed 10 ppm throughout the experiment and were lower than in the control poultry house. Aerobic mesophilic counts in air increased in both poultry houses in successive stages of production, but were lower in the house containing CaO_2 than in the control facility in weeks 3, 4, and 6. Calcium peroxide also reduced average yeast and mold counts in the experimental poultry house relative to control. The addition of CaO_2 at 2 g/m² litter did not compromise the birds' health and had a positive impact on performance.

In addition to technical devices and sanitary solutions, other measures are also introduced to minimize pollutant emissions from chicken houses. One of such measures relies on phytoremediation, namely the use of selected plants to accumulate and degrade polluting substances.

Sobczak et al. [56] and Domagalski et al. [57] analyzed the ability of selected greenhouse plants to reduce pollution levels in exhaust air from poultry houses. Sobczak et al. [56] demonstrated a decrease in the concentrations of carbon dioxide, ammonia, and dust when exhaust air was passed through an experimental greenhouse containing Indian shot (*Canna*) and silver grass (*Miscanthus*). Pollution levels were measured at the inlet and outlet of the greenhouse for 3 months to reveal a daily drop of 10% in CO_2 concentrations (30% during day time), a 40% drop in ammonia levels and a 14% drop in dust concentrations on average.

Domagalski et al. [57] investigated the deodorizing properties of Indian shot (*Canna × Generalis*) in a phytotron chamber for filtering exhaust air from poultry houses. The applied biofilter reduced total concentrations of odorous compounds by 20–30% and decreased ammonia levels by 29–41%.

5. Summary and conclusions

The results of the above studies demonstrate that poultry farms are significant reservoirs and emitters of microbiological and gaseous contaminants into the environment and that the type and concentrations of bioaerosols and gases produced in poultry farms are determined by various factors, including bird species, stocking density, season, time of day, stage of the production cycle, temperature, moisture content and the physicochemical parameters of litter, sampling site, ventilation efficiency, technical and process solutions, and farm management methods.

A microbiological analysis of bird facilities revealed that threshold concentrations of airborne bacteria and fungi recommended in the literature [27, 32] are often exceeded in practice. In cited studies, the lowest bacterial concentrations in a broiler house were determined at 3.1 $\log_{10}$ cfu/m^3; however, in the most cases, minimum value approximated the safe threshold for poultry houses (5.0 $\log_{10}$ cfu/m^3) already at the beginning of the production cycle. The highest concentrations of airborne bacteria were determined in hen houses at 8.3 $\log_{10}$ cfu/m^3. The proposed safe threshold for fungal concentrations of 3.3 $\log_{10}$ cfu/m^3 for poultry was also most often exceeded at the beginning of the production cycle, and fungal concentrations ranged from 2.7 (turkeys) to 5.9 $\log_{10}$ cfu/m^3 (broilers).

Poultry litter was an even more abundant source of microorganisms. In our study, the highest bacterial concentrations in hen house litter reached 9–10 $\log_{10}$ cfu/g, and the highest fungal concentrations reached 8 $\log_{10}$ cfu/g, but the above results cannot be compared with reference values due to an absence of normative threshold levels in the literature.

An IR spectroscopy analysis of chemical air pollution in a commercial turkey farm supported the determination of the type and concentrations of inorganic compounds (ammonia, carbon dioxide, nitric oxide, and phosphine) and VOC (sulfur and nitrogen compounds: nitriles, amines, and aldehydes; hydrocarbons: methane, dichloromethane, chloromethane, bromomethane, 1,3-butadiene). In most studies, volatile compounds in farm buildings are identified by gas chromatography. This method is characterized by high precision, but it is rarely used in practice because analyses have to be performed under laboratory conditions. A comparison of our findings with the results of chromatographic analyses indicates that FTIR is a practical method for evaluating gas contamination in field conditions because analyses can be conducted *in situ* with the use of a portable device, which eliminates the problems associated with sample collection and transport. IR spectroscopy supports the identification of aroma compounds even at very low concentrations.

Selected volatile compounds, which were also determined in our study of hen houses (e.g., ethanethiol, methanethiol, acrylonitrile), can be harmful at very low concentrations at the limit of detection of measuring devices equipped with electrochemical sensors for selective detection [58]. Due to analytical constraints, only general threshold limit values (TLV) have been determined for carbon dioxide, ammonia, and hydrogen sulfide at 1800–3000, 10–30, and 5–10 ppm, respectively, in housing facilities for juvenile and adult animals [18]. The relevant legal regulations set TLVs for NH_3, CO_2 and H_2S for calves and pigs, and NH_3 and CO_2 for chickens at 3000, 20, and 5 ppm, respectively. The regulations addressing other animal species, including turkeys, merely state that VOC concentrations should be kept at a safe level [59].

The growing number of protests staged by local communities against odor-producing animal farms, in particular animal production facilities situated inside the protective zone surrounding residential districts, has attracted researchers' attention to the odor-producing qualities of approximately 300 identified volatile compounds. The detection limit of many gases, including mercaptans, amines, sulfur compounds, and phenol derivatives, can be very low. Measures that effectively limit the production of odorous gas mixtures at the source require the identification of the highest number of components, even at very low concentrations. Analyses of trace amounts of toxic compounds are often burdened with error; therefore, the higher the number of replications and standardized measuring techniques, the greater the effectiveness of the proposed protective measures.

The EU climate and energy package places the Member States under the obligation to reduce their greenhouse gas emissions. Animal farms contribute to an increase in atmospheric concentrations of CO_2, CH_4, and NO_x. Farm emissions are determined based on standard formulas and computer simulations that do not account for hygiene standards. This approach could lead to unjust prosecution of farmers who take active steps to reduce pollution at the source. For this reason, gas concentrations and actual emissions from farm buildings characterized by different hygiene levels should be determined to effectively reduce atmospheric concentrations of pollutants.

In our study, the attempts to limit the concentrations of harmful gases and microbiological pollutants in farm buildings generated positive results. Total bacterial counts, including *Enterobacteriaceae* and *Staphylococcus* counts, were reduced in hen houses sprayed with essential oil solutions. Despite promising initial results, further analyses are needed to determine the effectiveness of essential oils in practice. Essential oils can be ineffective in small concentrations, and they can pose a health threat when applied excessively. The main advantage of essential oils is that they are effective antimicrobials, and microorganisms, which can acquire resistance to chemical substances, have not been found to develop a resistance to essential oils. The results of our study demonstrate that essential oil sprays could deliver even more satisfactory results in hen houses because their efficacy is determined by microbial species. The application of adsorbents also delivered promising results. Vermiculite and halloysite reduced VOC concentrations by 83 and 73%, respectively. Ammonia adsorption was determined at 15%. Despite the above, adsorbent efficiency was reduced over time as the volume of poultry droppings increased. The findings of other authors also indicate the positive results of biofilters, different additives to the litter and even phytoremediation in pollutants reduction in poultry houses.

Reliable criteria for evaluating poultry exposure to biological and chemical pollutants and the relevant reference values should be developed to maintain high poultry welfare standards in farms. For such criteria to be acceptable, they have to be carefully balanced to ensure that they deliver the highest level of poultry welfare and are achievable in practice with the involvement of the available methods.

Author details

Dorota Witkowska* and Janina Sowińska

*Address all correspondence to: dorota.witkowska@uwm.edu.pl

Department of Animal and Environmental Hygiene, Faculty of Animal Bioengineering, University of Warmia and Mazury in Olsztyn, Olsztyn, Poland

References

[1] Dutkiewicz J, Pomorski ZJH, Sitkowska J, Krysińska-Trawczyk E, Prażmo Z, Skórska C, Cholewa G, Wójtowicz H. Airborne microorganisms and endotoxin in animals house. Grana. 1994;33:84–90.

[2] Davis M, Morishita TY. Relative ammonia concentrations, dust concentrations and presence of *Salmonella* species and *Escherichia coli* inside and outside commercial layer facilities. Avian Diseases. 2005;49:30–35.

[3] Tymczyna L, Chmielowiec-Korzeniowska A, Drabik A. The effectiveness of various biofiltration substrates in removing bacteria, endotoxins and dust from ventilation system exhaust from a chicken hatchery. Poultry Science. 2007;86:2095–2100.

[4] Vučemilo M, Matković K, Vinković B, Jakšić S, Granić K, Mas N. The effect of animal age on air pollutant concentration in a broiler house. Czech Journal of Animal Science. 2007;52:170–174.

[5] Bakutis B, Monstviliene E, Januskeviciene G. Analyses of airborne contamination with bacteria, endotoxins and dust in livestock barns and poultry houses. Acta Veterinaria Brno. 2004;73:283–289.

[6] Wang Y, Chai T, Lu G, Quan C, Duan H, Yao M, Zucker BA, Schlenker G. Simultaneous detection of airborne aflatoxin, ochratoxin and zearalenone in a poultry house by immunoaffinity clean-up and high performance liquid chromatography. Environmental Research. 2008;107:139–144.

[7] Tymczyna L, Chmielowiec-Korzeniowska A, Drabik A, Skórska C, Sitkowska J, Cholewa G, Dutkiewicz J. Efficacy of a novel biofilter in hatchery sanitation: II. Removal of odorogenous pollutants. Annals of Agricultural and Environmental Medicine. 2007;14:151–157.

[8] Herbut E. The assessment of odors' emission from livestock. In: Szynkowska MI, Zwoździak J, editors. Modern problems of odours. WNT: Warsaw; 2010. p. 1–12.

[9] AL Homidan A, Robertson JF, Petchey AM. Review of the effect of ammonia and dust concentrations on broiler performance. World's Poultry Science Journal. 2003;59:340–349.

[10] Miles DM, Branton SL, Lott BD. Atmospheric ammonia is detrimental to the performance of modern commercial broilers. Poultry Science. 2004;83:1650–1654.

[11] Miles DM, Miller WW, Branton SL, Maslin WR, Lott BD. Ocular responses to ammonia in broiler chickens. Avian Diseases. 2006;50:45–49.

[12] Nahm KH. Evaluation of the nitrogen content in poultry manure. World's Poultry Science Journal. 2003;59:77–88.

[13] Guiziou F, Béline F. In situ measurement of ammonia and greenhouse gas emissions from broiler houses in France. Bioresource Technology. 2005;96:203–207.

[14] Wathes CM, Holden MR, Sneath RW, White RP, Phillips VR. Concentrations and emission rates of aerial ammonia, nitrous oxide, methane, carbon dioxide, dust and endotoxin in UK broiler and layer houses. British Poultry Science. 1997;38:14–28.

[15] Hayes ET, Curran TP, Dodd VA. Odour and ammonia emissions from intensive poultry units in Ireland. Bioresource Technology. 2006;97:933–939.

[16] Sówka I, editor. Methods of identification of odour gases emitted from industrial plants. Monographs No. 55. Publishing House of Wroclaw University of Technology: Wroclaw; 2011.

[17] Górny LR. Biohazards: standards, recommendations and threshold limit values. Podstawy i Metody Oceny Środowiska Pracy. 2004;3:17–39.

[18] Kołacz R, Dobrzański Z, editors. Livestock hygiene and welfare. Agricultural University in Wroclaw: Wroclaw; 2006. p. 76–81; 85–90.

[19] Witkowska D, Chorąży Ł, Mituniewicz T, Makowski T. Microbial contaminations of litter and air during broiler chickens rearing. Woda-Środowisko-Obszary Wiejskie. 2010;10:201–210.

[20] Baykov B, Stoyanov M. Microbial air pollution caused by intensive broiler chicken breeding. FEMS Microbiology Ecology. 1999;29:389–392.

[21] Wójcik A, Chorąży Ł, Mituniewicz T, Witkowska D, Iwańczuk-Czernik K, Sowińska J. Microbial air contamination in poultry houses in the summer and winter. Polish Journal of Environmental Studies. 2010;19:1045–1050.

[22] Lonc E, Plewa K. Comparison of indoor and outdoor bioaerosols in poultry farming. In: Anca Moldoveanu, editor. Advanced Topics in Environmental Health and Air Pollution Case Studies, InTech: Rijeka, Croatia; 2011. ISBN: 978-953-307-525-9, Available from: http://www.intechopen.com/books/advanced-topics-inenvironmental-health-and-air-pollution-case-studies/comparison-of-indoor-and-outdoor-bioaerosols-in-poultryfarming [Accessed: 2016-05-05]

[23] Bródka K, Kozajda A, Buczyńska A, Szadkowska-Stańczyk I. The variability of bacterial aerosol in poultry houses depending on selected factors. International Journal of Occupational Medicine and Environmental Health. 2012;25:281–293.

[24] Sowiak M, Bródka K, Kozajda A, Buczyńska A, Szadkowska-Stańczyk I. Fungal aerosol in the process of poultry breeding – quantitative and qualitative analysis. Medycyna Pracy. 2012;63:1–10.

[25] Plewa-Tutaj K, Pietras-Szewczyk M, Lonc E. Attempt to estimate spatial distribution of microbial air contamination on the territory and in proximity of a selected poultry farm. Ochrona Środowiska. 2014;36:21–28.

[26] Saleh M, Seedorf J, Hartung J. Inhalable and respirable dust, bacteria and endotoxins in the air of poultry houses [Internet]. 2007. Available from: http://citeseerx.ist.psu.edu/viewdoc/download?doi=10.1.1.566.8137&rep=rep1&type=pdf [Accessed: 2016-05-10].

[27] Lawniczek-Walczyk A, Gorny RL, Golofit-Szymczak M, Niesler A, Wlazlo A. Occupational exposure to airborne microorganisms, endotoxins and β-glucans in poultry houses at different stages of the production cycle. Annals of Agricultural and Environmental Medicine. 2013;20:259–268.

[28] Mituniewicz T, Sowińska J, Wójcik A, Iwańczuk-Czernik K, Witkowska D, Banaś J. Effect of disinfectants on physicochemical parameters of litter, microbiological quality of hen house air, health status and performance of broiler chickens. Polish Journal of Environmental Studies. 2008;17:745–750.

[29] Manafi M, Pirany N, Noor Ali M, Hedayati M, Khalaji S, Yari M. Experimental pathology of T-2 toxicosis and mycoplasma infection on performance and hepatic functions of broiler chickens. Poultry Science. 2015;94:1483–1492.

[30] Manafi M, Mohan K, Noor Ali M. Effect of ochratoxin A on coccidiosis-challenged broiler chicks. World Mycotoxin Journal. 2011;4:177–181.

[31] Seedorf J, Hartung J, Schroder M, Linkert KH, Phillips VR, Holden MR, Sneath RW, Short JL, White RP, Pedersen S, Takai H, Johnsen JO, Metz JHM, Groot Koerkamp PWG, Uenk GH, Wathes CM. Concentrations and emissions of airborne endotoxins and microorganisms in livestock buildings in Northern Europe. Journal of Agricultural Engineering Research. 1998;70:97–109.

[32] Krzysztofik B, editor. Microbiology of air. Publishing House of Warsaw University of Technology: Warsaw; 1992.

[33] Witkowska D. Volatile gas concentrations in turkey houses estimated by Fourier Transform Infrared Spectroscopy (FTIR). British Poultry Science. 2013;54:289–297.

[34] Witkowska D, Sowińska J. The effectiveness of peppermint and thyme essential oil mist in reducing bacterial contamination in broiler house. Poultry Science. 2013;92:2834–2843.

[35] Tymczyna L, Chmielowiec-Korzeniowska A. Reduction of odorous gas compound in biological treatment of ventilation air from layer house. Annals of Animal Science. 2003;3:389–397.

[36] Tymczyna L, Chmielowiec-Korzeniowska A, Saba L. Biological treatment of laying house air with open biofilter use. Polish Journal of Environmental Studies. 2004;13:425–428.

[37] Chmielowiec-Korzeniowska A, Tymczyna L, Drabik A, Malec H. Biofiltration of volatile organic compounds in the hatchery. Annals of Animal Science. 2005;5:371–278.

[38] Opaliński S, Korczyński M, Kołacz R, Dobrzański Z, Żmuda K. Application of selected alumonosilicates for ammonia adsorption. Przemysł Chemiczny. 2009;88:540–543.

[39] Opaliński S, Korczyński M, Szołtysik M, Dobrzański Z, Kołacz R. Applicationof aluminosilicates for mitigation of ammonia and volatile organic compound emissions from poultry manure. Open Chemistry. 2015;13:967–973.

[40] Oliveira MC, Almeida CV, Andrade DO, Rodrigues SMM. 2003. Dry matter content, pH and volatilized ammonia from poultry litter treated or not with different additives. Revista Brasileira de Zootecnia. 2003;32:951–954.

[41] Cook KL, Rothrock Jr., MJ, Eiteman MA, Lovanh N, Sistani K. Evaluation of nitrogen retention and microbial populations in poultry litter treated with chemical, biological or adsorbent amendments. Journal of Environmental Management. 2011;92:1760–1766.

[42] Korczyński M, Jankowski J, Witkowska D, Opaliński S, Szołtysik M, Kołacz R. Use of halloysite and vermiculite for deodorization of poultry fertilizer. Przemysł Chemiczny. 2013;92:1027–1031.

[43] Manafi M, Umakantha B, Narayana Swamy H, Mohan K. Evaluation of high-grade sodium bentonite on performance and immune status of broilers, fed ochratoxin and aflatoxin. World Mycotoxin Journal. 2009;2:435–440.

[44] Manafi M, Counteracting effect of high grade sodium bentonite during aflatoxicosis in broilers. Journal of Agricultural Science and Technology. 2012;14:539–547.

[45] Witkowska D, Sowińska J, Iwańczuk-Czernik K, Mituniewicz T, Wójcik A, Szarek J. The effect of a disinfection on the ammonia concentration on the surface of litter, air and the pathomorphological picture of kidneys and livers in broiler chickens. Archiv Tierzucht. 2006;49:249–256.

[46] Iwańczuk-Czernik K, Witkowska D, Sowińska J, Wójcik A, Mituniewicz T. The effect of a microbiological and a disinfecting preparation on the physical and chemical properties of litter and the results of broiler chicken breeding. Polish Journal of Natural Sciences. 2007;22:395–406.

[47] Witkowska D, Szarek J, Iwańczuk-Czernik K, Sowińska J, Mituniewicz T, Wójcik A, Babińska I. Effect of disinfecting litter on rearing performance and results of blood indices and internal organs of broiler chickens. Medycyna Weterynaryjna. 2007; 63:1115–1119.

[48] Cabuk M, Alcicek A, Bozkurt M, Akkan S. Effect of Yucca schidigera and natural zeolite on broiler performance. International Journal of Poultry Science. 2004;3:651–654.

[49] Ritz CW, Fairchild BD, Lacy MP. Implications of ammonia production and emissions from commercial poultry facilities: a review. Journal of Applied Poultry Research. 2004;13:684–692.

[50] Watson DW, Denning SS, Zurek L, Stringham SM, Elliott J. Effects of lime hydrate on the growth and development of darkling beetle, *Alphitobius diaperinus*. International Journal of Poultry Science. 2003;2:91–96.

[51] Bennett DD, Higgins SE, Moore RW, Beltran R, Caldwell DJ, Byrd JA, Hargis BM. Effects of lime on Salmonella enteritidis survival in vitro. The Journal of Applied Poultry Research. 2003;12:65–68.

[52] Bennett DD, Higgins SE, Moore RW, Byrd JA, Beltran R, Corsigli C, Caldwell DJ, Hargis BM. Effect of addition of hydrated lime to litter on recovery of selected bacteria and poultry performance. The Journal of Applied Poultry Research. 2005;14:721–727.

[53] Mituniewicz T. The effectiveness of calcium oxide (CaO) and calcium oxide-magnesium (CaOMgO) to liter in the rearing of broiler chickens [thesis]. Dissertations and Monographs: University of Warmia and Mazury in Olsztyn; 2012.

[54] Walawska B, Gluzińska J. Calcium peroxide as a source of active oxygen. Przemysł Chemiczny. 2006;85:877–879.

[55] Piotrowska J. Zoohygienic and productive parameters of the welfare of broiler chickens reared on liter with addition of calcium peroxide (CaO_2) [thesis]. University of Warmia and Mazury in Olsztyn; 2014.

[56] Sobczak J, Chmielowski A, Marek P, Rakowski A. Phytoremediation as a method of limiting pollutants contained in the air transmitted from a henhouse. Nauka Przyroda Technologie. 2011;5:1–14.

[57] Domagalski Z, Marek P, Sobczak J. Using the phytotron chamber to reduce the emission of offensive odour compounds from the poultry houses. Problems of Agricultural Engineering. 2012;76:127–136.

[58] Regulation of the Polish Ministry of Labor and Social Policy of 23 June 2014 on the highest threshold limit values of harmful substances in the work environment, Journal of Laws of 2014, item 817. Available from: http://isap.sejm.gov.pl [Accessed: 2016-05-05].

[59] Regulations of the Polish Ministry of Agriculture and Rural Development of 15 February 2010 and 28 June 2010 on the minimum rules of rearing for the livestock protection. Journal of Laws of 2010, No. 56, item 344; No. 116, item 778. Available from: http://isap.sejm.gov.pl [Accessed: 2016-05-05].

How to Control *Campylobacter* in Poultry Farms?: An Overview of the Main Strategies

Michel Federighi

Abstract

It is now recognized that *Campylobacter* is one of the main bacterial hazard involved in foodborne diseases around the world leading to an increasing number of gastrointestinal campylobacteriosis in humans. Also, it is known that this disease has a very high-social cost. According to researchers of Emerging Pathogens Institute (EPI) (University of Florida, the United States), the combination poultry/*Campylobacter* is the greatest cause of human campylobacteriosis. It is well known all around the world that intestinal carriage of *Campylobacter* is very large and frequent; it can be reached 100% of animal infected. Reducing this biological hazard can be exercised at different stage levels in the food chain. Intervention at the farm level by reducing colonization of the birds should be taken into account in the overall control strategy. This chapter gives an up-to-date overview of suggested on-farm control measures to reduce the prevalence and colonization of *Campylobacter* in poultry.

Keywords: *Campylobacter*, poultry, breeding, control strategies

1. Introduction

These days, the majority of human zoonotic microbial infections have a food origin (**Table 1**). Contamination of the food matrix can occur at all stages of the food production chain. In the search for causes of contamination, any stage of the production chain must not be neglected. This fact requires a global approach of problems and a good knowledge of the characteristics of the microorganisms involved. For the latter, the precise knowledge of their privileged reservoir and their potential ability to colonize other reservoirs will identify or clarify some contamination scenarios. Thus, it is known that the psychrotrophic nature of *Listeria monocytogenes* and their affinity for inert surfaces and biofilms found in food industries are promoting the contamination of the food matrix during the industrial stages of product processing [1].

Disease	Number of confirmed (a) human cases	Hospitalized cases	Reported deaths	Case-fatality (%)
Campylobacteriosis	236,851	18,303	25	0.01
Salmonellosis	88,715	9830	65	0.15
Yersiniosis	6625	442	5	0.13
VTEC infections	5955	930	7	0.20
Listeriosis	2161	812	210	15.0
Echinococcosis	801	122	1	0.51
Q-fever	777	NA	1	0.26
Brucellosis	347	142	0	0.00
Tularemia	480	92	0	0.00
Trichinellosis	319	150	2	0.84
West Nile fever (a)	77	48	7	13.7
Rabies	3	NA	2	100.0

(a) Exception made for West Nile fever where the total number of cases was included; NA: not applicable.

Table 1. Reported hospitalization, deaths, and case-fatality rates due to zoonoses in confirmed human cases in the EU, 2014.

Contamination of food appears as a necessary step to trigger disease in humans. In some cases, and for certain microorganisms, this phase must necessarily be followed by another phase involving a multiplication of microorganisms in food, concomitantly, or not, with a toxin synthesis. This second phase will allow microorganisms to reach sufficient numbers (minimum infectious dose) to cause disease in consumers. Thus, some microbial hazards should multiply in food (such as *Salmonella* or *L. monocytogenes*) and others not (*Campylobacter*, VTEC, for example).

Researchers from the Emerging Pathogens Institute (EPI) of the University of Florida in the United States have recently focused on infectious diseases of food origin. They estimated that 31 foodborne pathogens are responsible for 9.4 million cases of human infections each year in the United States, leading to 55,961 hospitalizations and 1351 deaths (http://www.epi.ufl.edu/?q=RankingTheRisks). Among all of these cases, 59% cases are associated with viruses, 39% cases with bacteria, and 2% cases by parasites. Among viruses, norovirus is involved in 58% of cases and for bacteria, *Campylobacter*, *Salmonella*, and *Clostridium perfringens* occupy the first three places of the ranking. The first two of this classification are confirmed in Europe where campylobacteriosis exceeds salmonellosis since 2008.

In fact, *Campylobacter* is considered the most abundant zoonotic agent in the European Union. Indeed, 190,566 cases of *Campylobacter* infections were reported in 2008, increasing annually to reach 236,851 in 2014 (**Table 1**). Salmonellosis are still in second place in this epidemiological study, with 88,175 cases in 2014 (against 131,468 cases reported in 2008), the *Listeria* bacteria are responsible for 2161 cases of infection in 2014 (against 1381 in 2008) with a high mortality rate (15%), especially among vulnerable people [2]. To illustrate the importance of *Campylobacter* infections in France, it is interesting to recall that the report of the "Institut de Veille Sanitaire" (InVS) in 2004 estimated the number of confirmed cases of campylobacteriosis in France to 21,652 cases, of which 17,322 cases were from food origin [3]. According to this report, 3516 cases had required hospitalization and 18 cases would have conducted to death. *Campylobacter* infections not only indirectly cause a high processing cost but also a high number of days of work stopping. For example, the treatment of campylobacteriosis in the UK is 465 € while it is 77 € in the Netherlands [4]. In Europe, the annual cost of campylobacteriosis treatment is of the order of 2.4 billion euros [5].

In Africa, the situation is most worrying. It is known that the first *Campylobacter* infection occurs early in life, by food or nonfood way. Indeed, children under 5 years are the most exposed at the campylobacteriosis [6]. So, Goualie et al. [7] reported an estimated incidence of campylobacteriosis between 40,000 and 60,000 for 100,000 children in developing countries of this continent. These numbers are increasing in most African countries. So, the means of control and prevention are more than necessary even though it has been shown that repeated infections in some children gave them a protection in front of the next infection [8].

Transmission by direct contact with reservoirs like pets, human being, or contaminated bathing water, although rare, should not be neglected. It can cause disease, especially for high-risk professions, namely: farmers, veterinarians, and slaughterhouse workers [9]. Notwithstanding, in most cases, transmission to humans is done indirectly by ingestion of water or food contaminated by certain species of *Campylobacter* called thermophilic (named *C. jejuni* ssp. *jejuni*, *C. coli*, *C. lari*, *C. upsaliensis*), naturally present in many farm animals (poultry, cattle, swine, etc.) [10]. In fact, researchers from the Emerging Pathogens Institute (EPI), mentioned above, were then interested in food vehicles of microbial hazards. Of the 14 most frequent pathogens involved in infectious disease of food origin and the 12 most consumed foods in the United States (or the 168 food-pathogen combinations studied), *Campylobacter*-chicken combination is the one that is causing the greatest number of cases followed by *Toxoplasma*-pork combination, *Listeria*-ready to eat meats combination, *Salmonella*-poultry, and *Listeria*-dairy products combinations [11]. In the United States, the campylobacteriosis is mainly sporadic cases of which it turns out that the contaminated chicken consumption is the cause of most cases. It seems that in Europe we can also highlight the *Campylobacter*-chicken combination since a study conducted in Belgium during the dioxin crisis, has shown that the number of campylobacteriosis decreased by 40% during the withdrawal period of the sale of poultry [12]. **Table 2** highlights the importance of chicken as *Campylobacter* vector compared with pork and beef meat (**Table 2**).

Meat	2006	2005	2004	2003	2002
Poultry	34.6%	30.5%	37.8%	35%	30.2%
Pork	0.7%	0.3%	1.6%	1.2%	1.4%
Beef	0.7%	0.9%	0.6%	0.3%	0.3%

Table 2. Comparison of different sources of *Campylobacter* contamination of poultry, pork and beef meat in Europe (years 2002–2006).

Some gestures made during the preparation of foods in the kitchen is often the cause of contamination transfers, including the use, for cutting the roasted poultry, of the board on which was cut or eviscerated raw poultry. Furthermore, studies have shown that the transfer of *Campylobacter* from chicken skin to kitchen work surfaces was possible at significant rate (from 0.05 to 36%), as well as to the hands of users (2.9–3.8%) [13]. It has, moreover, been shown that *Campylobacter* was able to survive for several hours on surfaces and stainless steel utensils, and sponges used for cleaning surfaces could also be sources of contamination [14].

Campylobacter jejuni is responsible for over 85% of campylobacteriosis. When it occurs, campylobacteriosis occurs typically after incubation for 24–72 h, by intestinal manifestations during a week. The most frequently described manifestation is acute gastroenteritis characterized by inflammation, severe abdominal pains in the periumbilical region, mucous diarrhea that can be bloody, accompanied sometimes by fever. It should be noted that the clinical manifestation is often less severe in developing countries where campylobacteriosis is manifested only by a significant watery diarrhea, this could be related to immune protection settling in individuals with frequent contact with *Campylobacter* [8].

This disease can be serious for certain populations or during postinfection complications, like Guillain-Barré syndrome or Miller-Fisher syndrome [15]. It seems that some serogroups of *C. jejuni*, as the serogroup O19 Penner, are particularly involved in this type of complications. Having a minimum growth temperature of 30°C, being intolerant to ambient oxygen and also being sensitive to technological stress (such as cold, heat, acidification, and drying), *Campylobacter jejuni* was always considered as a delicate and fragile organism [16]. Despite these nutritional requirements and the sensitivity to environmental stresses which prevent to grow and multiply outside the host or in food, *C. jejuni* is still able to survive and persist throughout the food production chain to cause campylobacteriosis constituting, in fact, a real paradox [17].

Although *Campylobacter* is found in the intestinal tracts of most red meat animals (cattle, pigs, and small ruminants) and pets (cats and dogs), the avian reservoir remains predominant due to the high carrier rate in animals and to the bacterial load per gram of feces, up to 10^7 CFU/g [13, 18]. Due to the large intestinal asymptomatic carriage in production animals, manure, slurry, soil, and water can also be reservoirs of *Campylobacter*. In fact, a study conducted in Italy showed that about 30% of water samples from rivers were contaminated with *Campylobacter jejuni* [19].

The colonization of the intestine of broilers by *Campylobacter* during rearing is responsible for contamination of carcasses after processing [20–22]. Worldwide, the average prevalence of

Campylobacter on poultry carcasses is about 60–80% [2, 23]. The carcass contamination occurs during the slaughtering process, even if certain operations are contaminating more than others. Thus, it is recognized that the contamination occurs more favorably during defeathering and evisceration, with feces leaking from the cloaca and the rupture of caeca, causing massive contamination by *Campylobacter* [24–26]. In addition, various transfers of contamination (or cross contamination) can intervene. Among these, there is a carcass to carcass contamination by contact and contamination transfer via vectors such as equipment and personnel, mostly [24].

All of these works clearly show that intestinal carriage of *Campylobater* by poultry is a key element of the transmission of this hazard to humans. Although beyond the breeding, control measures of this danger exist (good hygienic practices transformations, physical, and chemical treatments sanitizers), their effectiveness will be strengthened if the number of *Campylobacter* present at this stage is as low as possible. Therefore, the reduction or eradication of intestinal carriage in chickens is a strategic element of major importance for risk control *Campylobacter* in this sector especially as the contamination of poultry carcasses is proportional to the amount of *Campylobacter* present in caeca before slaughter [22]. In fact, a recent study of EFSA based on a quantitative microbiological assessment of risks (QMRA) evaluated the performance of implementation of interventions in primary production on reducing the risk of campylobacteriosis cases for the chicken consumers [27]. Following this study, the potential reduction of campylobacteriosis cases can be predicted after the application of breeding measures and is presented in **Table 3**.

Interventions in primary production		Reduction of campylobacteriosis cases
Improved hygiene/biosecurity		16%
Systematic use of screen fly in broiler houses (Denmark)		60%
Discontinued thinning		1.8–25%
Reduction of slaughter age	42 days	0–5%
	35 days	0.6–18%
	28 days	21–43%
Reduction colonization in cecal contents	1 log	48–83%
	2 logs	76–98%
	3 logs	90–100%
	6 logs	100%

Table 3. Effect of interventions in primary production on the reduction of human campylobacteriosis cases.

The study found that the most effective measures are those aimed at reducing the number of *Campylobacter* in caeca of chickens. However, other simpler measures also have a real impact, we now do an overview of all these measures.

2. Control *Campylobacter* in chicken farms

The few quantitative risk assessment studies available on the *Campylobacter*-chicken combination, conducted in different countries, often focus on the effectiveness of physical and chemical means of eliminating this zoonotic agent during the slaughter and the transformation process [28]. These methods have often proven their efficacy but are often oversized and sometimes poorly accepted by the consumer [10]. Preventive approaches such good hygiene practices and biosecurity now find some interest and may be a strategy to prevent the colonization of animals by *Campylobacter* and participate in the control of this zoonotic agent in the production of poultry meat. This interest is reinforced by the development of indirect measures, complementary of best practices, to reduce the intestinal number of *Campylobacter* in poultry. For some authors, these interventions would significantly reduce campylobacteriosis [29]. Thus, a Belgian study showed that the incidence of human campylobacteriosis in the country could be reduced by 32, 53, and 77% if the prevalence of broiler batches colonized by *Campylobacter* was reduced by 25, 50, or 75%, respectively [30].

2.1. Good hygienic practices and biosecurity

Thus, in addition to reducing the risk *Campylobacter* obtained at later stages of the human food chain, this reduction can also be achieved by the establishment of biosecurity measures at the breeding stage. These interventions are designed to protect a population of animals from the introduction of infectious agents transmissible like *Campylobacter*. In poultry, the biosecurity program includes all measures that must or may be taken to prevent the entry of this agent and changing the health status of the chicken population. These measures, collectively known as biosecurity, cover the hygienic practices during the rearing period. These include washing hands before entering in a poultry house, the use of different boots to enter each house, the cleaning, and disinfection of shoes before entering the room, a high level of hygienic water quality beverage.

Other measures such as cleaning and effective disinfection of poultry house between two batches of animals, as reducing the number of visits, as strict control of entry into the breeding of rodents, wild birds, and flying insects. Thus, studies in Denmark have shown that the use of mosquito nets preventing the entry of flying insects in the broiler house, potential vectors of *Campylobacter*, significantly reduced the contamination of poultry by *Campylobacter* during the seasonal peak (**Figure 1**) [31].

The application of all these measures greatly reduces the risk of *Campylobacter* infections. So, Gibbens et al. estimated that this application would lower the prevalence of *Campylobacter* in batches of chickens from 80 to less than 40% [32]. The respect of a good personal and clothing hygiene for staff and good measures of biosecurity, including the control of rodents and insects in two Dutch farms, reduced prevalence of *Campylobacter* in batches of chickens from two different farms of 34% in the first farm, and of 20% in the second [33].

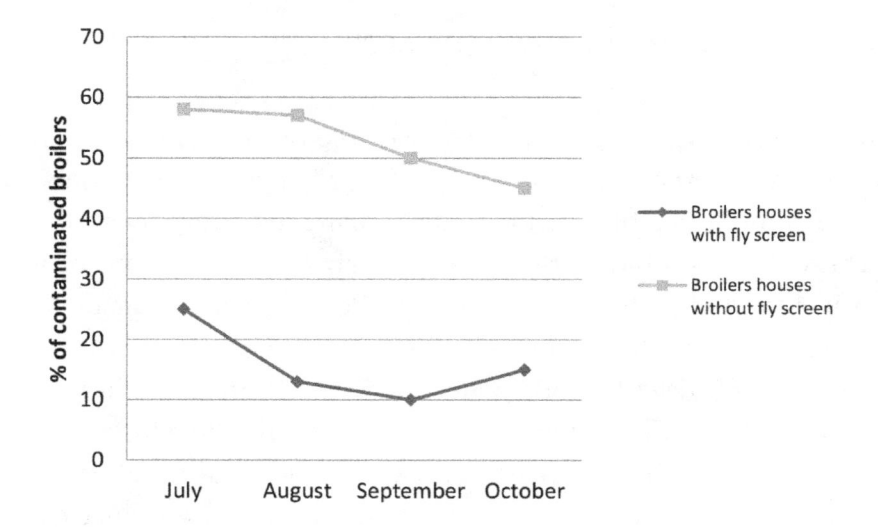

Figure 1. Effect of the use of fly screen in broilers houses on the percentage of contaminated broilers.

2.2. Treatment of drinking water

Another important factor is the quality of drinking water. Several studies have shown that poor quality water (untreated water from wells) may increase the transmission of *Campylobacter* in animals [34, 35]. The microbiological quality of drinking water should be monitored by the analysis and can be improved on the farm by techniques such as filtration, chlorination, ozonation, UV rays. For some authors, the impact of these interventions on *Campylobacter* infection is uncertain, but they point out that the absence of interventions may be worse [32, 36]. Studies by Byrd et al. showed that adding 0.44% (vol/vol) of lactic acid in drinking water prior to slaughter, has reduced the level of contamination of carcasses with *Campylobacter* [37]. Hilmarsson et al. showed that the addition of glycerol monocaprate (monocaprin) the last 3 days before slaughter, has reduced the number of *C. jejuni* in feces samples of chickens naturally or artificially infected [38].

2.3. Use of antimicrobial from vegetal origin

In addition to their application in drinking water, organic acids can also be used as additives in foods to reduce the prevalence of *Campylobacter* in poultry. Thus, 0.7% caprylic acid reduced colonization when used preventively on old chickens of 10 days and achieves a significant reduction of *C. jejuni* in broiler feces up to three to four decimal reductions [39]. In contrast, Van Deun et al. [40] observed that butyrate did not reduce colonization of caeca by *Campylobacter* in broiler chickens, but the addition of fatty acids short chain at a concentration of 1% reduces the risk of colonization of farms [41]. However, Hermans et al. [42] did not find any effect of these medium chain fatty acids (caproic, caprylic, and capric) on the number of *Campylobacter* in the caeca of broilers 28 days fed therewith 3 days before slaughter. Moreover, they observed that injection of a highly concentrated solution of sodium caprate directly into the caeca did not prevent colonization and did not reduce the caeca contents by *Campylobacter*. Thereafter, these authors showed that this ineffectiveness was explained by the presence of the intestinal mucus whom protecting *C. jejuni* in the caeca vis-a-vis of the bactericidal effect of organic acids observed *in vitro*. Conversely, another research group found a significant reduction (several logs) of *Campylobacter* in caeca of chickens when caprylic acid was given 3 days before slaughter [43]. In addition, another study showed that adding monocaprin to chicken feed the last 3 days before slaughter, resulted in a significant reduction of *C. jejuni* on feces samples of animals artificially or naturally infected, compared to controls [38]. These results, apparently contradictory, demonstrate the need to continue the investigations necessary to establish with greater certainty the effectiveness or lack of effectiveness of this strategy in the control of *Campylobacter* in chicken farms

2.4. Vaccination

The principle of vaccination of chickens against *Campylobacter* is to administer a product capable of inducing immunity directed specifically against this pathogen and confer immune memory enabling rapid activation (much shorter of lag period) of defenses in case of contamination. Vaccination could complement the use of biosecurity measures and other intervention strategies to reduce the level of contamination of poultry by *Campylobacter*. The development, production, and application of available vaccines would be beneficial to all parties involved and help enhance food safety and improve public health. Several studies on vaccination to reduce the sensitivity of broilers with *Campylobacter* colonization were performed. Khoury and Meinersmann [44] vaccinated chickens using a hybrid protein consisting of a portion of the FlaA *Campylobacter jejuni* flagellin (flagellar subunit) and the B subunit of heat-labile toxin (LT-B) of *Escherichia coli*. This results in a significant reduction of colonization of chickens by *Campylobacter*, and the production of specific antibodies against FlaA. Comparing with unvaccinated chicks, Rice et al. [45] have demonstrated a reduction of *Campylobacter* in chicks vaccinated orally with a combination of dead *Campylobacter jejuni* cells coupled to the heat labile toxin of *E. coli*.

More recent studies involving a larger number of animals were used to test the use of recombinant vaccines. Thus, 840 SPF chicks were used to evaluate the effectiveness of the vaccine derived from *Salmonella enterica* Typhimurium ΔaroA attenuated and expressing the immunogenic protein CJAA *C. jejuni*. Chicks who received the vaccine orally at the age of 1 day, then 2 weeks later, showed a reduction in fecal contamination by *C. jejuni* of 1.4log CFU/g, compared with unvaccinated chicks [46]. Layton et al. [47] used recombinant vaccines attenuated from Salmonella expressing three peptide epitopes of protein of *Campylobacter* (Omp18/CJAD protein, CJAA, and Cj0420 (ACE393)). These three vaccines were administered orally to chicks on the day 1, then 21 days later. The vaccinated chicks were inoculated with *C. jejuni*. Eleven days after the inoculation, an increase in IgG and IgA antibodies specifically against *C. jejuni* was observed and also a reduction in the number of *C. jejuni* in the ileum. Vaccination was most effective when the vectored vaccine expressing the epitope of the Omp18/CJAD protein was administered to chicks, with a considerable reduction of *C. jejuni* in the chicken intestine (4.8 decimal reductions of *C. jejuni* in the ileum) compared with unvaccinated controls and those vaccinated only with the vector (*Salmonella* 13A) or negative control (**Figure 2**).

These studies are promising and probably mean that a possible vaccination strategy for *Campylobacter* reduction is possible. They still face a lack of information on the immune system of the chick that such hinders the development of an attenuated vaccine expressing the linear peptide epitope *Campylobacter* (Omp18/CJA). In addition, advances in functional genomics *Campylobacter* suggest that other proteins of this agent could be excellent candidates for testing for future vaccines.

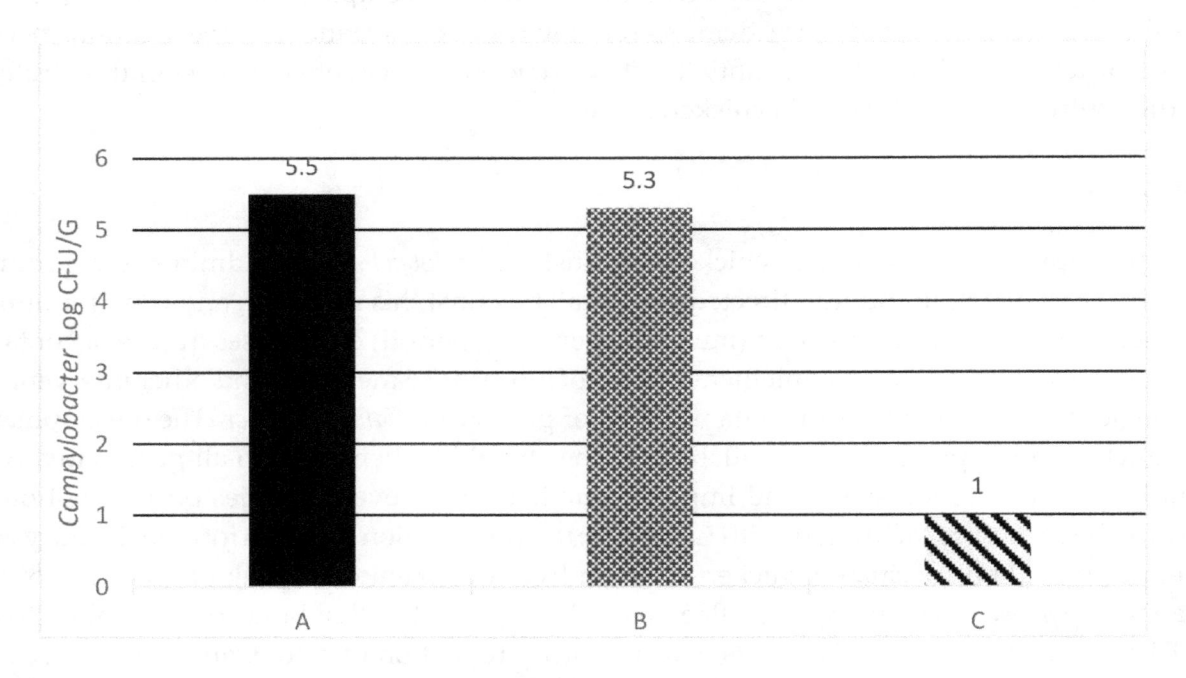

Figure 2. Enumeration of *Campylobacter* (log CFU/cecal content) on 1 day orally vaccinated chicken. Group A: salted water vaccinated; group B: *Salmonella* 13A vaccinated, and group C: *Campylobacter* CJ0113 vaccinated.

2.5. Use of phages

The lytic activity of bacteriophages can be used as a strategy to reduce the colonization of chickens with *Campylobacter*. Phages usually have a very narrow spectrum of activity, and they do not interact with other bacterial species in the intestinal flora. Phages bind and penetrate into bacterial cells by protein receptors and multiply within the cytoplasm until the death of the bacteria. At this time, the lysis of the bacteria permits the release of new bacteriophages.

Loc Carrillo et al. [48] and Wagenaar et al. [49] have shown three decimal reductions of *Campylobacter* in caeca of chickens that received the bacteriophage, compared with negative control. However, this reduction is not stable, they observed a reduction of only 1log/g 5 days later. Similarly, El-Shibiny et al. [50] observed an immediate reduction of $2log_{10}(CFU/g)$ of *Campylobacter* in caeca 2 days after ingestion of phages. Then, the number of *Campylobacter* in caeca returns to the original level a few days later, reversing the improvement achieved. These results show that this strategy is more a short-term therapeutic strategy than a preventive long term one. It could be very interesting if, for example, the treatment is taken only 2–3 days before slaughter: resistance to bacteriophages then not have time to reverse the reduction of *Campylobacter* obtained by the treatment. In this matter, other studies have also shown that administration of phages in feed is more effective than oral gavage [37]. In these conditions of administration, use of phages few days before slaughter appears to be an excellent strategy for the reduction of *Campylobacter* in poultry, but the diversity of *Campylobacter* protein receptors requires a large diversity of phages, which also increases the complexity of this strategy.

2.6. Use of prebiotics and probiotics

Probiotics are defined as "live microorganisms which when administered in adequate amounts confer a benefit to human health." Prebiotics are generally oligosaccharides (fructo-oligosaccharide (FOS), galacto-oligosaccharides (GOS)) or polysaccharides such as inulin. These escape digestion in the small intestine and have a beneficial effect on the health of their host by stimulating the growth and/or activity of bacteria of the genera *Lactobacillus* and *Bifidobacterium*, naturally present in the colon or administered as probiotics.

The use of prebiotics and probiotics is a strategy that has been studied by several research teams in order to reduce the colonization of chickens by *Campylobacter jejuni*.

In 1997, Morishita et al. [51] used on 1-day-old chicks, a probiotic mixture containing *Lactobacillus acidophilus* and *Enterococcus faecium*. The chicks were randomly divided into two groups, one group was treated with the probiotic cocktail during the first 3 days of culture while the second batch received distilled water instead of the probiotic cocktail. Six hours after the first oral administration of probiotics, the number of *C. jejuni* in feces of the chicks was determined and was performed until the slaughter. The results showed a 70% reduction in the concentration of *C. jejuni* in chickens on day 3 and a 27% reduction for broilers at slaughter, compared to the control group. For Schoeni and Wong, administration of a mixture of different bacteria (*Citrobacter diversus*, *Klebsiella pneumonia*, and *E. coli*) appears to be effective in preventing or reducing the colonization of chickens with *Campylobacter* [52]. This protection has been strengthened by mannose which was given as a prebiotic. In 2000, Chang and Chen [53] tested on *C. jejuni*, in an *in vitro* model of the digestive tract of chicken, the effect of a mixed culture of *Lactobacillus acidophilus*, *L. fermentum*, *L. crispatus*, and *L. brevis* in a feed wherein mannose has been added, showing a inhibitory effect. Similarly, Baurhoo et al. [54] observed a significant reduction of *C. jejuni* in naturally contaminated caeca of broiler, who received a diet containing mannanoligosaccharide as a prebiotic.

Again, this is very promising works, that requires further study in order to decide definitively on their use. They also have the merit of bringing forward an interesting and ongoing concept named "microbial solution for microbial problems."

2.7. Genetic selection of chicken

Selective breeding of resistant lines of chickens to *Campylobacter* colonization is a particularly modern intervention strategy to reduce *Campylobacter* problem in the poultry industry.

In 2005, Boyd et al. [55] have shown that the selection of chicken lines genetically resistant to *Campylobacter* germ, significantly reduces this risk in poultry. In their study, Boyd et al. [55] have inoculated 1-day-old chicks of different inbred lines with 10^7–10^8 CFU of *C. jejuni* or *C. jejuni* 81–176 ^{14}N and measured bacterial colonization levels of chickens over a period of 2–3 weeks. They have always been a difference of a factor 10–100 from four inbred lines in the number of *C. jejuni* present in chicken caeca between the four inbred chicken lines. The biggest difference was for the N line, which presented relatively high levels of *Campylobacter*, and the line 61, which had a relatively small number of bacteria. Among the four lines studied, the major histocompatibility complex does not appear to be a major factor in determining the resistance. The difference in the number of bacteria in fecal samples was observed after 24 h after inoculation and was still present at the end of the experiment. This work revealed that the difference in the number of bacteria was inherited in a consistent manner with the resistance (low number of bacteria), controlled by a single autosomal dominant locus. These data suggest that it may be possible to identify the responsible genes. Indeed, the recent knowledge of the whole sequence of the chicken genome has identified the genes involved in susceptibility to *Campylobacter* colonization [56]. These observations led to the suggestion that selective breeding could be used to select chickens resistant to *Campylobacter* colonization.

2.8. Use of bacteriocins

The use of antimicrobial peptides could be an interesting biological intervention strategy to reduce colonization of poultry by *Campylobacter* [57–61]. These studies highlight the ability of bacteriocins produced by lactic acid bacteria, such as *Lactobacillus salivarius* NRRL B-30514, *Enterococcus faecium* E50-52 E760, *Lactobacillus salivarius*, and *L. salivarius* SMXD51 1077 (NRRL B-50053)) to inhibit the growth of *Campylobacter jejuni*. In 2005, Stern et al. [62] studied the effect of the bacteriocin "SRCAM 602" produced by *P. polymixa* NRRL B-30509 on cecal colonization by *Campylobacter* of chicks artificially inoculated with 10^8 CFU of *C. jejuni* from day 1. These animals, colonized by *Campylobacter*, received from day +7 to day +10 a feed containing the bacteriocin purified (250 mg/kg). For the chicks that received during 3 days, this diet the number of *Campylobacter* in caeca was very low and undetectable (<$2\log_{10}$ CFU/g), while the control animals showed high cecal colonization by *Campylobacter* (10^6–10^8 log CFU/g) [62]. In 2008, Line et al. [57] have found similar results following administration of the bacteriocin "Enterococcine E-760" in broiler naturally infected with *Campylobacter*. Moreover, Svetoch et al. [63] administered 10.8 mg/chicken (oral gavage) of bacteriocin "E 50-52" produced by *E. faecium* NRRL B-30746 3 days before slaughter. The results showed a significant reduction of *Campylobacter* in the gut, greater than 10^5 CFU/g of feces.

Stern et al. [64] have studied the effect of the bacteriocin OR 7, produced by *L. salivarius* NRRL B-30514, in encapsulated form in a concentration of 250 mg/kg on eight groups of contaminated chicks by *C. jejuni*. In three of eight groups of chickens, there was no *Campylobacter* colonization and, in the other five groups, the level of contamination remained very low (10–100 CFU/g) compared with controls. By cons, *L. salivarius* NRRL B-30514 and *Paenibacillus polymyxa* NRRL-B-30509 showed no effect on colonization of artificially infected chicks by *C. jejuni* [64]. Finally, Svetoch et al. [63] showed that treatment with the bacteriocin L-1077, produced by *L. salivarius* strain NRRL B-50053, chickens inoculated with *C. jejuni* and *Salmonella* Enteritidis provides more than four decimal reductions in the number of bacteria per gram of cecal contents, compared with controls. Moreover, the presence of these bacteria in the liver and spleen of the animals is very greatly reduced.

3. Conclusion

Campylobacter is today a leading cause of foodborne diseases, all around the world. It is also a paradox for microbiologists, who see a contradiction between the apparent physiological fragility, its small genome and its obvious ability to survive outside its main habitat (digestive tract of birds) and to reach its main target (i.e., the consumer). Moreover, this impression is reinforced by the fact that the organism does not grow in foods and that his number would tend to decrease during processing operations, rather than increase. In fact, intestinal carriage of *Campylobacter* by animals becomes a key element of the contamination of the consumer and a series of strategies have been developed to reduce intestinal carriage in the past 15 years. Today, despite all the efforts and progress made, there is still no miracle solution but a set of interventions strategies, each with their advantages and disadvantages. The use of bacteriocins and bacteriophages is very promising because their implementation is simple: they can be easily administered with water or feed. However, their potential use requires further research work on their long-term effectiveness. In addition, the successful application of these methods as well as probiotics, prebiotics and even vaccination may be affected by genomic instability of *C. jejuni* [44] which may affect the effectiveness of these strategies in the long term. Finally, the reduction of *Campylobacter* contamination pressure of animals at the stage of livestock should be based on a strong base of Good Hygienic Practices and biosecurity, reinforced by targeted interventions selected on the criteria of efficiency, practicality, and cost in front of the type of poultry production.

Author details

Michel Federighi

Address all correspondence to: michel.federighi@oniris-nantes.fr

Oniris, Food Quality and Hygiene, Nantes, France

References

[1] Ribeiro da Silva M, Federighi M. Mint: seafood products as a source of listeriosis: myth or reality? Advanced Research in Food Science. 2005; 4, 73–78.

[2] EFSA. Mint: analysis of the baseline survey on the prevalence of *Campylobacter* in broiler batches and of *Campylobacter* and *Salmonella* on broiler carcasses in the EU, 2008, part A: *Campylobacter* and *Salmonella* prevalence estimates. EFSA Journal, 2010; 8 (3), 1503. doi:10.2903/j.efsa.2010.1503.

[3] INVS. Morbidity and mortality due to foodborne infections in France. Report of the French Institute of Public Health. 2004; pp. 185.

[4] Roberts JA, Cumberland P, Sockett PN, Wheeler J, Rodrigues LC, Sethi D, Roderick PJ. Mint: the study of infectious intestinal disease in England: socio-economic impact. Epidemiology and Infection. 2003; 130, 1–11.

[5] EFSA. Mint: scientific opinion on *Campylobacter* in broiler meat production: control options and performance objectives and/or targets at different stages of the food chain. EFSA Journal. 2011; 9 (4), 2105. doi:10.2903/j.efsa.2011.2105.

[6] Oberhelman R, Taylor D. *Campylobacter* infections in developing countries. In: Nachamkin I, Blaser M, editors. *Campylobacter*, 3rd ed. Washington: ASM Press; 2000. p. 139–153.

[7] Goualie GB, Karou G, S Bakayoko, KJ Coulibaly, Coulibaly KE, Niamke SL, Mr. Dosso Mint: Prevalence of Campylobacter in chickens sold in the markets of Abidjan: Pilot study in the municipality of Adjamé 2005 African Journal of Health and Animal Production. 2010; 8 (S), 31–34.

[8] Lastovica A, Skirrow M. Clinical significance of *Campylobacter* and related species other than *Campylobacter jejun*i and *C. coli*. In: Nachamkin I, Blaser M, editors. *Campylobacter*, 3rd ed. Washington: ASM Press; 2000. p. 89–120.

[9] Garénaux A Ritz-Bricaud M, Mr. Federighi Mint: Campylobacter and food safety: risk analysis, risk assessment and risk management. Bulletin of the Veterinary Academy of France, 2005; 158, 377-383.

[10] Federighi M, C Magras, Pilet MF. Campylobacter. In: Federighi M, editor. Food bacteriology, Compendium of food hygiene, Paris, Economica, 2005, p. 145-172.

[11] Scallan E, Hoekstra R, Angulo F, Tauxe R, Widdowson MA, Roy S, Jones J, Griffin P. Mint: foodborne illness acquired in the United States—major pathogens. Emerging and Infectious Disease. 2011; 17 (1), 7–15.

[12] Vellinga A, Loock VID. Mint: the dioxin crisis as experiment to determine poultry-related *Campylobacter enteritis*. Emerging and Infectious Disease. 2002; 8, 19–22.

[13] Salvat G, Chemaly M, Denis M, Robinault C, Huneau A, Le Bouquin S, Fravalo P. Mint: evolution des risques sanitaires: *Campylobacter* et Salmonelles. Sciences des Aliments. 2008, 28 (4–5), 285–292.

[14] Kusumaningrum HD, Riboldi G, Hazeleger WC, Beumer RR. Mint: survival of foodborne pathogens on stainless steel surfaces and cross-contamination to foods. International Journal of Food Microbiology. 2003; 85, 227–236. doi:10.1016/S0168-1605(02)00540-8; doi:10.1016/S0168-1605%2802%2900540-8#doilink.

[15] Zilbauer M, Dorrell N, Elmi A, Lindley KJ, Schüller S, Jones HE, Klein NJ, Núňez G, Wren BW, Bajaj-Elliott M. Mint: a major role for intestinal epithelial nucleotide oligomerization domain 1 (NOD1) in eliciting host bactericidal immune responses to

Campylobacter jejuni. Cellular Microbiology. 2007; 9, 2404–2416. DOI: 10.1111/j.1462-5822.2007.00969.x.

[16] Martinez-Rodriguez S, Mackey BM. Mint: factors affecting the pressure resistance of some *Campylobacter* species. Letters in Applied Microbiology. 2005; 41, 321–326. DOI: 10.1111/j.1472-765X.2005.01768.x.

[17] Messaoudi S, Federighi M. Campylobacter infections: epidemiology, diagnosis, control and prevention. In: Bertucci BA, editor. Campylobacter Infection: Epidemiology, Clinical Management and Prevention. New York: Novascience Publishers; 2015. p. 29–54.

[18] Newell D. Mint: the ecology of *Campylobacter jejuni* in avian and human hosts and in the environment. International Journal of Infectious Disease. 2002; 6, (Suppl. 3), S16–S21. doi:10.1016/S1201-9712(02)90179-7; doi:10.1016/S1201-9712%2802%2990179-7#doilink.

[19] Yan SS, Pendrak ML, Foley SL, Powers JH. Mint: *Campylobacter* infection and Guillain-Barre syndrome: public health concerns from a microbial food safety perspective. Clinical Applied Immunology Review. 2005; 5, 285–305. doi:10.1016/j.cair.2005.08.001; doi:10.1016/j.cair.2005.08.001#doilink.

[20] Herman L, Heyndrickx M, Grijspeerdt K, Vandekerchove D, Rollier I, De Zutter L. Mint: routes for *Campylobacter* contamination of poultry meat: epidemiological study from hatchery to slaughterhouse. Epidemiology and Infection. 2003; 131, 1169–1180. doi.org/10.1017/S0950268803001183; doi:10.1017/S0950268803001183#_blank.

[21] Rasschaert G, Houf K, Van Hende J, De Zutter L. Mint: *Campylobacter* contamination during poultry Slaughter in Belgium. Journal of Food Protection. 2006; 69, 27–33.

[22] Reich F, Atanassova V, Haunhorst E, Klein GN. Mint: the effects of Campylobacter numbers in caeca on the contamination of broiler carcasses with *Campylobacter*. International Journal of Food Microbiology. 2008; 127, 116–120. doi:10.1016/j.ijfoodmicro.2008.06.018; doi:10.1016/j.ijfoodmicro.2008.06.018#doilink.

[23] Suzuki H, Yamamoto S. Mint: *Campylobacter* contamination in retail poultry meats and by-products in Japan: a literature survey. Food Control. 2009; 20, 531–537. doi:10.1016/j.foodcont.2008.08.016; doi:10.1016/j.foodcont.2008.08.016#doilink.

[24] Allen VM, Weaver H, Ridley AM, Harris JA, Sharma M, Emery J, Sparks N, Lewis M, Edge S. Mint: sources and spread of thermophilic *Campylobacter* spp. during partial depopulation of broiler chicken flocks. Journal of Food Protection. 2008; 71, 264–270.

[25] Berrang ME, Buhr RJ, Cason JA, Dickens JA. Mint: broiler carcass contamination with *Campylobacter* from feces during defeathering. Journal of Food Protection. 2001; 64, 2063–2066.

[26] Boysen L, Rosenquist H. Mint: reduction of thermotolerant *Campylobacter* species on broiler carcasses following physical decontamination at slaughter. Journal of Food Protection. 2009; 72, 497–502.

[27] Romero-Barrios P, Hempen M, Messens W, Stella P, Hugas M. Mint: quantitative microbiological risk assessment (QMRA) of food-borne zoonoses at the European level. Food Control, 2013; 29(1), 341–349. doi:10.1016/j.foodcont.2012.05.043; doi:10.1016/j.foodcont.2012.05.043#doilink.

[28] Rosenquist H, Sommer HM, Nielsen NL, Christensen BB. Mint: the effect of slaughter operations on the contamination of chicken carcasses with thermotolerant *Campylobacter*. International Journal of Food Microbiology. 2006; 108, 226–232. doi:10.1016/j.ijfoodmicro.2005.12.007; doi:10.1016/j.ijfoodmicro.2005.12.007#doilink.

[29] Lin J. Mint: novel approaches for *Campylobacter* control in poultry. Foodborne Pathogens and Diseases. 2009; 6, 755–765. doi:10.1089/fpd.2008.0247.

[30] Messens W, Hartnett E, Gellynck X, Viaene J, Halet D, Herman L, Grijspeerdt K. Quantitative risk assessment of human campylobacteriosis through the consumption of chicken meat in Belgium. In: Proceedings of the XVIII European Symposium on the Quality of Poultry Meat and XII European Symposium on the Quality of Eggs and Egg Products, 2–5 September 2007; Prague. p. 167–168.

[31] Hald B, Sommer HM, Skovgard H. Mint: use of fly screens to reduce *Campylobacter* spp., introduction in broiler houses. Emerging and Infectious Diseases. 2007; 13, 1951–1953.

[32] Gibbens JC, Pascoe SJS, Evans SJ, Davies RH, Sayers AR. Mint: a trial of biosecurity as a means to control *Campylobacter* infection of broiler chickens. Preventive Veterinary Medicine. 2001; 48, 85–99. doi:10.1016/S0167-5877(00)00189-6; doi:10.1016/S0167-5877%2800%2900189-6#doilink.

[33] Van De Giessen AW, Tilburg JJ, Ritmeester WS, Van Der Plas J. Mint: reduction of *Campylobacter* infections in broiler flocks by application of hygiene measures. Epidemiology Infection. 1998; 121, 57–66.

[34] Lyngstad TM, Jonsson ME, Hofshagen M, Heier BT. Mint: risk factors associated with the presence of *Campylobacter* species in Norwegian broiler flocks. Poultry Science. 2008, 87, 1987–1994. doi: 10.3382/ps.2008-00132.

[35] Sparks NHC. Mint: the role of the water supply system in the infection and control of *Campylobacter* in chicken. World Poultry Science Journal. 2009; 65, 459–474. doi.org/10.1017/S0043933909000324; doi:10.1017/S0043933909000324#_blank.

[36] Mohyla P, Bilgili SF, Oyarzabal OA, Warf CC, Kemp GK. Mint: application of acidified sodium chlorite in the drinking water to control *Salmonella serotype Typhimurium* and *Campylobacter jejuni* in commercial broilers. Journal of Applied Poultry Research. 2007; 16, 45–51. doi: 10.1093/japr/16.1.45.

[37] Byrd JA, Hargis BM, Caldwell DJ, Bailey RH, Herron KL, Mac Reynolds JL, Brewer RL, Anderson RC, Bischoff KM, Callaway TR, Kubena LF. Mint: effect of lactic acid

administration in the drinking water during preslaughter feed withdrawal on *Salmonella* and *Campylobacter* contamination of broilers. Poultry Science. 2001; 80, 278–283. doi: 10.1093/ps/80.3.278.

[38] Hilmarsson H, Thormar H, Thrainsson JH, Gunnarsson E, Dadadottir S. Mint: effect of glycerol monocaprate (monocaprin) on broiler chickens: an attempt at reducing intestinal *Campylobacter* infection. Poultry Science. 2006; 85, 588–592. doi: 10.1093/ps/85.4.588.

[39] De Los Santos FS, Donoghue AM, Venkitanarayanan K, Metcalf JH, Reyes-Herrera I, Dirain ML, Aguiar VF, Blore PJ, Donoghue DJ. Mint: the natural feed additive caprylic acid decreases *Campylobacter jejuni* colonization in market-aged broiler chickens. Poultry Science. 2009; 88, 61–64. doi: 10.3382/ps.2008-00228.

[40] Van Deun K, Haesebrouck F, Van Immerseel F, Ducatelle R, Pasmans F. Mint: short-chain fatty acids and l-lactate as feed additives to control *Campylobacter jejuni* infections in broilers. Avian Pathology. 2008; 37, 379–383. doi: 10.1080/03079450802216603.

[41] Gerwe TV, Bouma A, Klinkenberg D, Wagenaar JA, Jacobs-Reitsma WF, Stegeman A. Mint: medium chain fatty acid feed supplementation reduces the probability of *Campylobacter jejuni* colonization in broilers. Veterinary Microbiology. 2010; 143, 314–318. doi:10.1016/j.vetmic.2009.11.029; doi:10.1016/j.vetmic.2009.11.029#doilink.

[42] Hermans D, Martel A, Van Deun K, Verlinden M, Van Immerseel F, Garmyn A, Messens W, Heyndrickx M, Haesebrouck F, Pasmans F. Mint: intestinal mucus protects *Campylobacter jejuni* in the ceca of colonized broiler chickens against the bactericidal effects of medium-chain fatty acids. Poultry Science. 2010; 89, 1144–1155. doi:10.3382/ps.2010-00717.

[43] De Los Santos FS, Hume M, Venkitanarayanan K, Donoghue AM, Hanning I, Slavik MF, Aguiar VF, Metcalf JH, Reyes-Herrera I, Blore PJ, Donoghue DJ. Mint: caprylic acid reduces enteric *Campylobacter* colonization in market-aged broiler chickens but does not appear to alter cecal microbial populations. Journal of Food Protection. 2010; 73, 251–257.

[44] Khoury C, Meinersmann RJ. Mint: a genetic hybrid of the *Campylobacter jejuni* flaA gene with LT-B of *Escherichia coli* and assessment of the efficacy of the hybrid protein as an oral chicken vaccine. Avian Diseases. 1995; 39 (4), 812–820. doi: 10.2307/1592418.

[45] Rice BE, Rollins DM, Mallinson ET, Carr L, Joseph SW. Mint: *Campylobacter jejuni* in broiler chickens: colonization and humoral immunity following oral vaccination and experimental infection. Vaccine, 1997; 15, 1922–1932. doi:10.1016/S0264-410X(97)00126-6; doi:10.1016/S0264-410X %2897%2900126-6#doilink.

[46] Buckley AM, Wang J, Hudson DL, Grant AJ, Jones MA, Maskell DJ, Stevens MP. Mint: evaluation of live-attenuated *Salmonella* vaccines expressing *Campylobacter* antigens for control of *C. jejuni* in poultry. Vaccine. 2010; 28, 1094–1105.

[47] Layton SL, Morgan MJ, Cole K, Kwon YM, Donoghue DJ, Hargis BM, Pumford NR. Mint: evaluation of *Salmonella*-vectored *Campylobacter* peptide epitopes for reduction of *Campylobacter jejuni* in broiler chickens. Clinical Vaccine Immunology. 2011; 18, 449–454. doi: 10.1128/CVI.00379-10.

[48] Loc Carrillo C, Atterbury RJ, El-Shibiny A, Connerton PL, Dillon E, Scott A, Connerton IF. Mint: bacteriophage therapy to reduce *Campylobacter jejuni* colonization of broiler chickens. Applied and Environmental Microbiology. 2005; 71, 6554–6563. doi: 10.1128/AEM.71.11.6554-6563.

[49] Wagenaar JA, Bergen MAPV, Mueller MA, Wassenaar TM, Carlton RM. Mint: phage therapy reduces *Campylobacter jejuni* colonization in broilers. Veterinary Microbiology. 2005; 109, 275–283. doi:10.1016/j.vetmic.2005.06.002 doi:10.1016/j.vetmic.2005.06.002# doilink.

[50] El-Shibiny A, Scott A, Timms A, Metawea Y, Connerton P, Connerton I. Mint: application of a group II *Campylobacter* bacteriophage to reduce strains of *Campylobacter jejuni* and *Campylobacter coli* colonizing broiler chickens. Journal of Food Protection. 2009; 72, 733–740.

[51] Morishita TY, Aye PP, Harr BS, Cobb CW, Clifford JR. Mint: Evaluation of an avian-specific probiotic to reduce the colonization and shedding of *Campylobacter jejuni* in broilers. Avian Diseases. 1997; 41, 850–855. doi: 10.2307/1592338.

[52] Schoeni JL, Wong AC. Mint: Inhibition of *Campylobacter jejuni* colonization in chicks by defined competitive exclusion bacteria. Applied and Environmental Microbiology. 1994, 60, 1191–1197.

[53] Chang MH, Chen TC. Mint: reduction of *Campylobacter jejuni* in a simulated chicken digestive tract by *Lactobacilli* cultures. Journal of Food Protection. 2000; 63, 1594–1597.

[54] Baurhoo B, Ferket PR, Zhao X. Mint: effects of diets containing different concentrations of mannanoligosaccharide or antibiotics on growth performance, intestinal development, cecal and litter microbial populations, and carcass parameters of broilers. Poultry Science. 2009; 88, 2262–2272. doi: 10.3382/ps.2008-00562.

[55] Boyd Y, Herbert EG, Marston KL, Jones MA, Barrow PA. Mint: host genes affect intestinal colonisation of newly hatched chickens by *Campylobacter jejuni*. Immunogenetics. 2005; 57, 248–253. doi:10.1007/s00251-005-0790-6.

[56] Kaiser P, Howell MM, Fife M, Sadeyen JR, Salmon N, Rothwell L, Young J, Poh TY, Stevens M, Smith J, Burt D, Swaggerty C, Kogut M. Mint: towards the selection of chickens resistant to *Salmonella* and *Campylobacter* infections. Bulletin des Membres de l'Académie Royale de Médecine de Belgique. 2009; 164, 17–25; discussion 25-16.

[57] Line JE, Svetoch EA, Eruslanov BV, Perelygin VV, Mitsevich EV, Mitsevich IP, Levchuk VP, Svetoch OE, Seal BS, Siragusa GR, Stern NJ. Mint: isolation and purification of enterocin E-760 with broad antimicrobial activity against gram-positive and gram-

negative bacteria. Antimicrobial Agents Chemotherapy. 2008; 52, 1094–1100. doi: 10.1128/AAC.01569-06.

[58] Messaoudi S, Kergourlay G, Rossero A, Ferchichi M, Prevost H, Drider D, Manai M, Dousset X. Mint: identification of *Lactobacilli* residing in chicken ceca with antagonism against *Campylobacter*. International Microbiology. 2011; 14, 103–110.

[59] Stern NJ, Svetoch EA, Eruslanov BV, Perelygin VV, Mitsevich EV, Mitsevich IP, Pokhilenko VD, Levchuk VP, Svetoch OE, Seal BS. Mint: isolation of a *Lactobacillus salivarius* strain and purification of its bacteriocin, which is inhibitory to *Campylobacter jejuni* in the chicken gastrointestinal system. Antimicrobial Agents Chemotherapy. 2006; 50, 3111–3116. doi: 10.1128/AAC.00259-06.

[60] Svetoch EA, Eruslanov BV, Perelygin VV, Mitsevich EV, Mitsevich IP, Borzenkov VN, Levchuk VP, Svetoch OE, Kovalev YN, Stepanshin YG, Siragusa GR, Seal BS, Stern NJ. Mint: diverse antimicrobial killing by *Enterococcus faecium* E 50-52 bacteriocin. Journal of Agricultural Food Chemistry. 2008; 56, 1942–1948. doi: 10.1021/jf073284g.

[61] Svetoch EA, Stern NJ. Mint: bacteriocins to control *Campylobacter* spp. in poultry — a review. Poultry Science. 2010; 89, 1763–1768. doi:10.3382/ps.2010-00659.

[62] Stern NJ, Svetoch EA, Eruslanov BV, Kovalev YN, Volodina LI, Perelygin VV, Mitsevich EV, Mitsevich IP, Levchuk VP. Mint: *Paenibacillus polymyxa* purified bacteriocin to control *Campylobacter jejuni* in chickens. Journal of Food Protection. 2005; 68, 1450–1453.

[63] Svetoch EA, Eruslanov BV, Levchuk VP, Perelygin VV, Mitsevich EV, Mitsevich IP, Stepanshin J, Dyatlov I, Seal BS, Stern NJ. Mint: isolation of *Lactobacillus salivarius* 1077 (NRRL B-50053) and characterization of its bacteriocin, including the antimicrobial activity spectrum. Applied and Environmental Microbiology. 2011; 77, 2749–2754. doi: 10.1128/AEM.02481-10.

[64] Stern NJ, Eruslanov BV, Pokhilenko VD, Kovalev YN, Volodina LL, Perelygin VV, Mitsevich EV, Mitsevich IP, Borzenkov VN, Levchuk VP, Svetoch OE, Stepanshin YG, Svetoch EA. Mint: bacteriocins reduce *Campylobacter jejuni* colonization while bacteria producing bacteriocins are ineffective. Microbial Ecology and Health Diseases. 2008; 20, 74–79. doi:10.1080/08910600802030196.

Biofilms of *Salmonella* and *Campylobacter* in the Poultry Industry

Daise A. Rossi, Roberta T. Melo,
Eliane P. Mendonça and Guilherme P. Monteiro

Abstract

Biofilm is characterized by a bacterial population firmly adhered to a surface involved by a self-produced matrix of extracellular polymeric substance. These communities provide longer survival and resistance to adverse conditions such as presence of antibiotics and disinfectants. Various foodborne microorganisms are capable of forming such structures, including *Salmonella* and *Campylobacter*, which are the major contaminants at the poultry industry. This biomass can affect the water transport system and pipes, and once the agent is established at the industry, it can form biofilms in any processing area. There are intrinsic and extrinsic mechanisms, and also molecular aspects involved in the biofilm formation. The adoption of several strategies may exhibit effectiveness to prevent the cell adhesion, such as the use of surfaces resistant to biofilm formation. In case of preexisting biofilms, there are physical, chemical, and biological methods used to control and eliminate them. Nanotechnology has emerged as another effective measure as nanometals affect the essential activities of microorganisms. These findings highlight the difficulty in controlling biofilms, due to the strategies used by these agents to adapt and survive in sessile form, causing recurring contamination throughout the poultry chain production, deterioration in the final product and infections in the human host.

Keywords: prevention, control, public health, microbial adhesion, anti-biofilm agents

1. Introduction

The industry destined to food production has a great challenge to maintain the safety of the products to be marketed. Among these challenges, there are the failures that may occur during the cleaning process, which may favor the permanence of microorganisms that are able to form biofilms. Several microorganisms transmitted by food are able to form these structures, among them stands out gender representatives *Salmonella* and *Campylobacter*, which are major contaminants of poultry products and the agents that represent the highest risk of foodborne infections to humans [1].

Biofilms are defined as a process of bacterial cells adhesion to a living or inert surface. These cells clump together forming bacterial communities, which are surrounded by a polymer matrix composed mainly by polysaccharides, as well as proteins and nucleic acids [2, 3]. This extracellular matrix promotes the biofilm protection, inhibiting access of biocidal agents, concentrating nutrients, and preventing dehydration [4].

Typically, biofilms consist of microorganisms in mixed cultures under symbiotic conditions, which are considered more resistant to chemical agents commonly used for cleaning and sanitizing, as well as to other harsh conditions such as refrigeration, acidity, saltiness, and antibiotics [5–7].

Since *Salmonella* and *Campylobacter* are present in the industry environment, they can form biofilms on produced food, and also in processing areas, such as walls, floors, pipes, and drains, and in contact surfaces, such as stainless steel, aluminum, nylon, rubber, plastic, polystyrene, and glass [8, 9].

The broiler slaughter industry generates residue rich in protein and lipids, which are deposited on surfaces [10], favoring the formation of biofilms of these pathogens responsible for frequent public health problems. Thus, these bacteria end up becoming a potential source of contamination within the industry that can be transferred to food or to their packaging, becoming a constant threat of recontamination [11].

Biofilms cause economic losses due to food spoilage and damage to equipment by biocorrosion, and also for damages caused in humans arising from foodborne infections [12].

The infections by *Salmonella* and *Campylobacter* have similar characteristics, with presence of diarrhea, abdominal cramps, and fever between 12 and 72 hours after infection. The illness usually lasts 4–7 days, and in most people, the recovery takes place spontaneously, without treatment. In some cases, diarrhea can be more serious with the need for hospitalization. In these patients, the microorganisms can spread from the intestines to the bloodstream and then to other body sites, with the risk of death. In cases of campylobacteriosis, patients may develop Guillain-Barré syndrome (GBS), which causes flaccid paralysis and risk of death from respiratory insufficiency [13].

There are intrinsic and extrinsic mechanisms involved in the formation of these communities, such as the material, type, and shape of the surface, the electric charge, hydrophobicity, and hydrodynamics. There are also different characteristics that determine the maturity of the biofilm, including the environment, mobility, growth rate, the capacity of cell signaling, and the production of extracellular polymer matrix [14]. Besides these, the molecular aspects are also extremely important, as the presence of *luxS* gene and activation of the *quorum-sensing* system in *Campylobacter* and genes encoding extracellular matrix components, Curli fimbriae and cellulose in *Salmonella* [15, 16].

The biofilm maturation allows the development of a primitive homeostasis and circulatory system, with exchange of genetic material and metabolic cooperation coordinated by *quorum-sensing*. This system is the mechanism used by bacteria to withstand the changes in their environment and therefore use specific strategies that allow adaptation to environmental stress [17].

About the problems involved in the presence of biofilms, preventing their development and their elimination represents greater security to the produced food and consumer. The adoption of several strategies may exhibit effectiveness in eliminating the use of more resistant surfaces for biofilm formation [18]. In the prevention and in cases of previous formed biofilms, physical, chemical, and biological methods can be used, being the combination of the three methods considered most effective [19]. Nanotechnology has emerged as another alternative as nanometals affect the essential activities of *Campylobacter* and *Salmonella* [20].

This data reinforces the difficulty in *Salmonella* and *Campylobacter* control in sessile forms within the chain production of food, especially in poultry products industry.

2. Characteristics of biofilms

Biofilms are formed by an aggregation of bacteria that adhere to each other and secrete extracellular polymeric substances (EPS), as a pellicle formed in the air-liquid interface. These structures are established in response to various environmental conditions and can be developed by multiple signaling strategies. They are usually composed of a mixture of different species of bacteria [21].

There are five stages that make up the cycle of formation of these structures: (a) free phase, (b) fixing the surface, (c) microcolony, (d) macrocolony, and (e) dispersion (**Figure 1**).

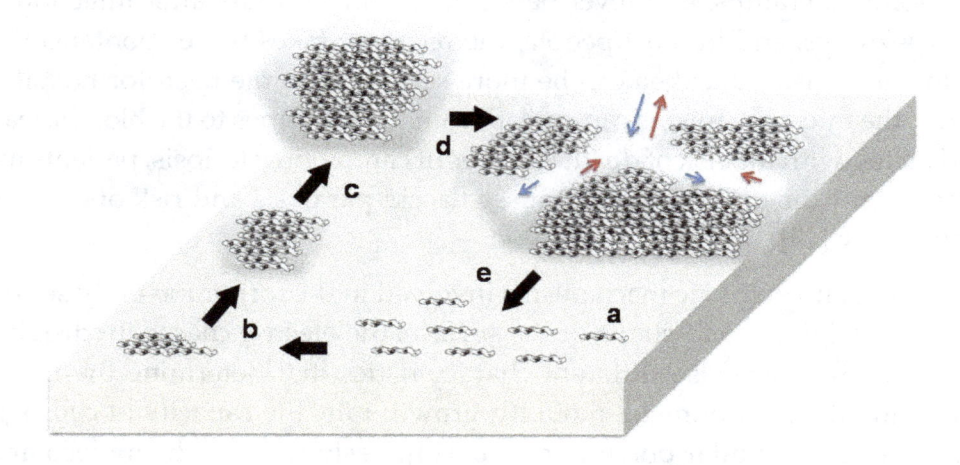

Figure 1. Stages of biofilm formation cycle in *Campylobacter*: (a) planktonic phase, (b) adhesion to the surface, (c) microcolony, (d) macrocolony, and (e) dispersion. The arrows indicate the path taken by nutrients (blue) and excreta (red) inside the tubules formed in the mature biofilm.

The first stage is characterized by reversible fixing facilitated by flagellar motility that allows the range to the surface. This connection is of low intensity and allows full movement of the bacterium, but also allows it to be easily removed by cleaning processes [22]. This weak initial interaction of the bacteria with the substrate involves hydrophobic interactions, electrostatic forces, and van der Waals force, which determine the adhesion between the bacterial cell and the surface [23].

The second stage is called irreversible fixation, moment when gradual increasing in the bond strength occurs by means of continuous production of exopolymers and adhesins. At this stage, the cell removal from the surface requires the action of mechanical force such as scraping, or by chemical treatment. The most important components for this period are exopolymers, adhesins, and DNA (extracellular DNA with structural function) [24]. This phase involves stronger dipole-dipole, hydrogen bonds, hydrophobic interactions, covalent, and ionic bonds [23].

In *Campylobacter*, there is an exponential increase in the expression of genes related to flagellar motility (*flaA* and *flaB*), and adhesion such as *cadF* and *peb4*. Bacterial motility plays a key role in the migration and starts the formation of microcolonies mediated by *quorum-sensing* mechanism [25, 26].

The formation of microcolonies is detected in the third stage, which occurs approximately after 2 hours, to *Salmonella*, and 4 hours to *Campylobacter*, and after the fixing step, with a diameter ranging from 0.5 to 2 mm. The cells become sessile and the *quorum-sensing* mechanism is activated upregulating the expression of the *luxS* gene, engaged in self-induction of biofilm formation. At that time, extrinsic factors act directly on the adhesion capacity, such as environmental conditions (aerobic, anaerobic, and microaerophilic), the type of available nutrients and the attachment surface (stainless steel, glass, and polypropylene) [21].

The maturation step is accompanied by the formation of macrocolonies resulting from microbial growth and recruitment of other environmental microorganisms. At this time, diffusion through the matrix of exopolymers is slower than the cellular metabolism and the resulting chemical gradients create microniches. Inside them, the death of bacterial cells is evident in the central region, permitting the formation of cavities where motility is possible for planktonic forms. *In vitro* that stage lasts for up to 3 days, if exchanged for fresh media and is accompanied by the formation of channels that allow the exchange of nutrients and excretions [24, 27]. **Figure 2** shows a schematic structure of bacterial biofilm *in vitro*.

The last phase is the dispersion of planktonic forms for formation of new biofilms. This process is done passively, independently of oxygen concentration [24, 28].

In human and animal hosts, the *C. jejuni* and *Salmonella* are capable of forming microcolonies on the intestinal epithelium within a few hours after infection. Biofilm formation in the intestinal absorption surfaces prevents normal functions in the ileal mucosa and can contribute to the symptoms of the disease. The ability of adhesion and chemotaxis to mucin, which comprises intestinal mucus, allows the formation of "bacterial blankets" inside and under these layers of mucus after 3–4 hours of exposure [22].

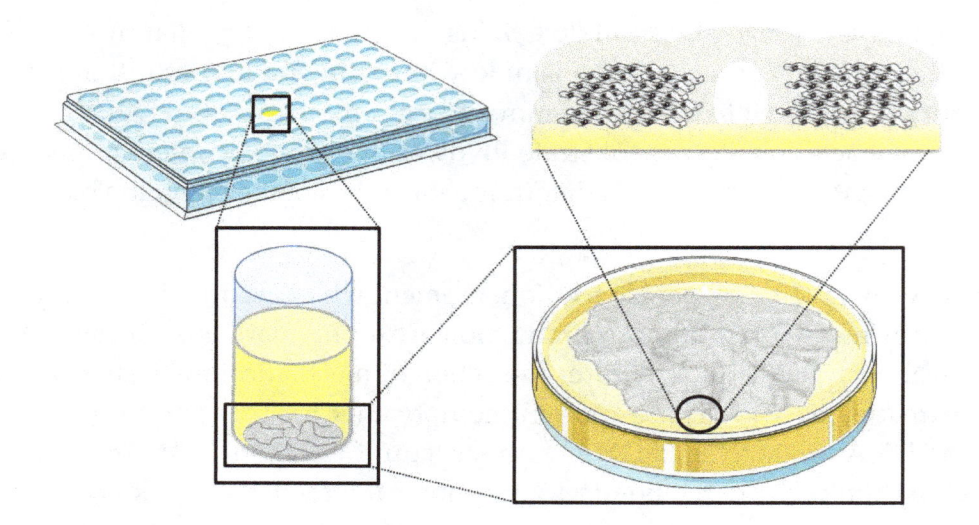

Figure 2. *In vitro* analysis of bacterial biofilms formation. The testing protocol is based on 10^7 UFC.mL^{-1} in specific culture media (TSB broth, Mueller Hinton broth, Bolton broth, Brucella broth, and Chicken juice broth) incubated for 24 hours (*Salmonella*) or 48 hours (*Campylobacter*) in plates of 96 wells at a temperature of 37°C under aerobic (*Salmonella*) or microaerophilic (*Campylobacter*) conditions. Subsequently, there is the formation of a bacterial mass on the bottom of the wells, which characterize the structure of a biofilm. After about 96 hours of incubation, the formation of substances transport channels is already established and becomes visible in electronic confocal and scan microscopies.

Outside the host, sessile form of *Campylobacter* and *Salmonella* allows better survival under stressful conditions of temperature, oxygen, and nutrients in abiotic environments, in different food matrices, and especially, in chicken meat and in the presence of antibacterial agents. Thus, its spread through the food chain becomes easier. The reasons that allow this survival success are multifactorial, but include especially the reduction of metabolic activity and decreasing of the adsorbent action of extracellular polymer matrix, which reduces the amount of antimicrobial and sanitizing agent required to interact with the biofilm cells, and specific factors expressed by the cells in the biofilm, such as efflux pumps [8, 29].

The behavior of these pathogens in mixed and single cultures differs significantly. Biofilms formed by different species of bacteria, such as those found in food industries, represent a substantial risk, since they can protect each other during the application of chemical agents [30]. It is true that in mixed cultures with *Pseudomonas* spp., *Staphylococcus* spp., *Escherichia coli*, *Bacillus* sp., and *Enterococcus faecalis* survival and persistence of *Campylobacter* is quite evident. In addition, structural changes and in metabolic activity of microcolonies can be observed. Additional examples are the commensalism of *C. jejuni* with *Pseudomonas aeruginosa*, which promotes the increase of tolerance to environmental oxygen concentrations by *C. jejuni* and the increase of *Campylobacter* survival in water biofilms composed of different species of protozoa [31–33].

3. Biofilms in poultry production

Salmonella and *Campylobacter* are microorganisms that are usually contaminants of chicken flocks. Some risk factors are the key to raising levels of infection, including inadequate hygiene measures, negligence of biosecurity standards, the presence of other animals on the premises, and others. However, in addition to environmental issues, there is also the persistence of the microorganism in the form of biofilm on the premises [34].

The absence of water is lethal to the growth of microorganisms, but the existence of a minimum supply of water within the farm may be sufficient for the establishment of biofilm. Drinking fountains are more conducive to bacterial attachment, being the portions covered with rubber with greater biomass, because of the facility to adhere [1].

The control of biofilm formation in the water distribution systems contribute significantly to improve the health of birds, minimizing the need of antibiotic treatments. All chemicals that are added to the drinking water of the poultry, such as dietary supplements and medicines, settle in biofilms and then may spread and generate residues even after its use has been completed. In that way, cleaning of the drinking fountains of poultry is a practice of extreme importance to be adopted in the farms, to ensure that the chickens receive good quality water [35]. The treatment of water supplied in poultry has gained increasing popularity in Europe [36], and today it is an essential measure to control pathogens in the poultry production chain.

In order to reduce microbial contamination the producers treat the water that will be destined to the birds with various chemicals such as: chlorine, chlorine dioxide, organic acids, peracetic acid, and hydrogen peroxide. However, these substances have action only under appropriate conditions of temperature and pH [37] and for a limited period of time, so it should be repeated periodically.

Some producers use chemical control methods in water intended for poultry only once or twice a week, and others limit this kind of processing to the end of the production cycle, i.e., there is a lack of standardization and proper methodology that undertake control efficiency and reduction of pathogens in the water, and potentiates the transmission of pathogens by water [37].

Studies have shown that one of the most common sources of contamination are from biofilms developed inside the water supply pipe, where a variety of microorganisms proliferate surrounded by mud, and are adhered to surface and continuously release planktonic cells in water [38].

Ventilation systems (coolers) are also a favorable area for microbial aggregation, especially in situations where there is a preexisting biofilm [1].

Higher biofilm formation rates are observed at the chicken processing steps because of large amount of moisture in the environment. Several critical points are identified during production, such as plastic curtains, mats, scalding tanks, chiller, and stainless steel tools [18].

A research with *Salmonella* spp. strain isolated from a poultry slaughterhouse in Brazil, determined that the canvas was the most suitable material for the induction of growth of the matrix by this microorganism, followed by polystyrene, and lastly, stainless steel [9].

The fact that the chicken reaches the abattoir harboring pathogens such as *Salmonella* and *Campylobacter*, increase the chances of contamination of equipment during processing.

Campylobacter genotyping studies clearly show the similarity of strains present in the intestine of poultry, and in environmental samples of the final product [34, 39].

The survival of *Campylobacter* in the chicken's skin is another form of adaptation of the microorganism. This agent is able to fix into the deep crevices of the skin and follicles feather of the bird. These recesses provide ideal conditions for bacteria to adhere, colonize, form biofilms, and remain protected in the carcass, even at low temperatures [40].

Thus, it must consider that proper control of cleaning methods used in the poultry production system is of paramount importance and should include strict compliance with the established biosecurity protocols. This applies mainly to difficult decontamination environments such as feed mills, agricultural environments, and farms [41, 42].

4. Intrinsic mechanisms of *Campylobacter*

For the biofilm establishment, both environmental variations and the microorganism itself correlate with genes that are expressed by bacteria in sessile form.

Table 1 illustrates some of the molecular mechanisms involved directly in biofilm formation and the link between this mechanism and the flagellar apparatus.

The several genes that encode different flagellar proteins clearly show the necessity of flagella on biofilm formation in *C. jejuni*, since they are overexpressed in this form of growth. The absence of *flaA, flaB, flaC, flhA,* or *fliS* results in the formation of weak biofilms. There are many flagella functions for biofilm formation and are linked to the motility, adhesion, and secretion of substances, suggesting its importance in the early stages of formation of sessile structure [43].

The lack of genes involved in flagella expression activation, as well as those that are involved in chemotaxis (*cheA, cheY,* and *cetB*) also reduces the self-bonding [44].

The biofilm matrix is composed basically by exopolysaccharides (EPS), proteins and DNA and thus the regulatory genes of these molecules production and the availability of nutrients define quantitative variations in this composition [21].

In addition to the flagella, the genes that decisively affect biofilm formation in *Campylobacter* are associated with the envelope/cell surface. Mutations that alter the expression and the protein secretion of surface also impair biofilm formation [45, 46].

The stress response generates metabolic changes in the expression of nearly the totality of the bacterial genome. Both, increasing the response to stress protein expression and decreasing of metabolic activity express the conditions of sessile microorganism and contributes to tolerance to the harsh conditions that bacteria are submit in biofilms. Iron uptake and stress response proteins are highly expressed in biofilms. Metabolic protein expression will vary, depending on the need, suggesting that distinct changes in metabolism mark the transition between modes of planktonic and sessile growth [47, 48].

	Gene	Proteic product	Biofilm	Motility
Flagella adhesion	*flaA*	Larger flagellin	+	+
	flaB	Flagellin	+	=
	flaC	Flagellin homologous, adhesion	+	=
	fiA	Flagellar sigma factor	+	=
	fibA	Flagellar proteic secretor apparatus	+	+
	flaG	Flagellin homologus	+	=
	fliS	Flagellar chaperone	+	+
	cadF	Adhesin	+	+
Metabolism	*pta*	Acetyltransferase phosphate	+	+
	ackA	Acetate kinase	+	+
	phoX	Alkaline phosphatase	–	=
Regulation/stress response	*csr A*	*RNA-binding* regulatory protein	+	+
	spoT	Binfunctional synthetase II	–	=
	ppk1	Poly-P kinase 1	–	=
	ppk2	Poly-P kinase 2	–	=
	Cj1556	Transcriptional regulator	+	+
	luxS	Autoinducer-2 synthase	+	+
	sodB	Oxidative stress regulator	+	=
	dnaJ	Thermic stress regulator	+	+
	cbrA	Metabolic growth regulator	+	=
	htrA	Osmotic shock regulator	+	=
	p19	Iron uptake	+	=
Proteic envelop/secretion	*pgp1*	Carboxypeptidase	+	+
	peb4	Peptidyl-prolyl isomerase	+	+
	tatC	Arginine transporter	+	+

+, Increase in phenotype;
–, Decrease in phenotype;
=, Remains the same phenotype.
Source: [50] with modifications.

Table 1. Phenotypic classification of biofilm formation capacity and flagellar motility according to the presence of specific genes.

5. Intrinsic mechanisms of *Salmonella*

Salmonella is equipped with external cellular components, such as flagella, fimbriae, and pili, which play an important role in the accession process of the cells to surfaces, environmental persistence, in biofilm formation, and in colonization and cell invasion [49].

Even in the case of intrinsic mechanisms of *Salmonella*, Hamilton et al. [50] observed the existence of several genes and proteins involved in bacterial attachment, motility, detection, and response to oxygen availability, transport and response to stress, being differently expressed in biofilms than the planktonic cells.

Among the most important genes associated with biofilm formation, there is the *csg*, responsible for coding the Curli fimbriae, described to be highly aggregative and play an important role in biofilm formation by *Salmonella*, to promote the initial interaction between the cell and the surface and the subsequent cell-cell interaction [51]. CsgD previously referred to AgfD, is the main unit of control and integration for biofilm formation in *Salmonella*, regulating the expression of specific matrix components associated to biofilm [52]. The *csg*D gene acts in addition to the regulation of fimbriae, activation of cellulose, determining the formation of a dense extracellular matrix. This gene activates the transcription of *adrA* gene, which stimulates the production of a protein that acts on the enzymatic activity of the cellulose biosynthesis. Thus, it forms a highly hydrophobic compressed network and composing rigid lining cells in the extracellular matrix of *Salmonella* biofilms [53].

6. Extrinsic factors linked to biofilms

About the environmental conditions, there are many factors that determine the production, development, and maintenance of biofilms, including the pH, temperature, type of material and the surface roughness, the presence of organic and inorganic compounds, the condition of dynamic flow, osmotic pressure, oxygen concentration, concentration and bioavailability of nutrients, and the presence of antimicrobial agents in the medium. This is due to the fact that different environmental conditions will generate different responses in gene regulations of the bacteria and thus the behavior of biofilms [54].

The sessile mode of growth is enhanced in low quantity of nutrient conditions. This fact is noted by elevated LPS production in the matrix. An example is related to the use of excessive nutritive media, such as Bolton, Brucella, and Brain Heart Infusion broths, *in vitro*, that are less prone to biofilm formation in *C. jejuni* and *Salmonella*, than Mueller Hinton, Chicken juice, and Tryptone Soya broth, less rich in nutrients. The carbon deficiency, nitrogen and phosphorus, is related to increasing formation of biofilms [27, 55].

The role of temperature in the formation of biofilms is more complex, varies among species and the changes are related to other environmental conditions [56].

For *Campylobacter*, unlike what happens in other enterobacteria including *Salmonella*, a dynamic flow condition does not lead to a better fixation. Thus, *in vitro* assays are carried out in static conditions, since the agitation does not allow connection of the microorganisms to form biofilms. In contradictory ways, in mixed cultures of biofilms, agitation rates can be high and equal to 2.5 mL/min [57].

The physicochemical properties of the surface exert a strong influence on the adherence of microorganisms. In general, the bacteria adhere more easily to the hydrophobic surfaces like plastics, than hydrophilic surfaces such as glass or metal [58].

The osmotic stress inhibits biofilm formation and leads to dispersion of the existing structure. The addition of NaCl (sodium chloride), glucose, or sucrose significantly decreases the formation of biofilm on *C. jejuni*, being these induced to transition to coccoid morphology [55].

The effect of oxygen tension in biofilm formation of *C. jejuni* appears to vary widely among strains. The oxygen seems to promote the early stages of biofilm formation, which occurs more rapidly in the first 24 hours. However, after 48 hours there are no significant differences in aerobic and microaerophilic conditions [25, 56].

7. The importance of the *quorum-sensing*

The *quorum-sensing* mechanisms directly influence the formation of biofilms. This communication of numerous bacteria via small signal molecules directs bacterial population and regulates the expression of virulence genes, toxin production, motility, chemotaxis, and biofilm formation, which can contribute to the adaptation and bacterial colonization [59].

Molecules of *quorum-sensing* AI-1 (autoinducer 1) are signaling to mediate specific intraspecies communication. AI-2 mediates communication intra and inter-particular species of Gram-positive and Gram-negative bacteria [59].

For *C. jejuni*, it has shown that AI-2 in the supernatant level increases during the exponential growth phase and decreases during entry into stationary phase. This molecule is produced by *Campylobacter* in different food matrices at various temperatures and corresponds to the expression of *luxS* gene, which is overexpressed in cultures of Chicken juice at low temperatures [60].

Both *quorum-sensing* mechanisms (AI-1 and AI-2) contribute to the regulation of biofilm formation against the growth of planktonic cells, which in turn promotes bacterial colonization, persistence under adverse conditions and expression of virulence factors. Thus, the detailed investigation of AI-mediated mechanisms could serve as a new tool to be used in therapeutic applications or reducing the amount of human pathogens in the transmission path for the termination of bacterial communication [61].

8. Biofilm control

Given the biofilms resistance of *Salmonella* and *Campylobacter* to disinfectants and antibiotics, it is important to evaluate and develop alternative strategies to prevent their formation.

The equipment design and the choice of the materials and coatings used in the food industry are extremely important in preventing biofilm formation. This is because even adopting the most effective cleaning and sanitizing programs, it is not possible to compensate for problems caused by faulty equipment, which have inaccessible corners, cracks, crevices, valves, and joints, which are vulnerable points for biofilm accumulation [62].

The use of well-designed equipment associated with the adoption of effective hygiene measures allow the removal of unwanted material from surfaces, including microorganisms, foreign materials, and residues from cleaning products [63, 64].

New technologies for detecting the presence of biofilms have been developed in order to control the colonization of surfaces by bacteria and identify the early stages of biofilm formation and development [65]. A research performed by Ref. [65] developed a mechatronic sensor to surface capable of providing various information such as the presence of biofilms in the early stages, presence of cleaning products in the surface, and differentiation of the type of cleaning employed (biological or chemical).

Once the biofilm is already established, it should be emphasized cleaning processes using mechanical action, which is one of the main measures for their elimination or control [66], because the friction acts on the matrix disruption, exposing deeper layers and making the microorganisms more accessible.

Generally, disinfectants do not penetrate the biofilm matrix after an inefficient cleaning procedure and, therefore, does not destroy all the biofilm cells [64], reaching only the outer layers. Cleaning is the first step and very important to improve the sanitation of equipment and facilities [67]. It is important to remove effectively the food wastes that may contain microorganisms or promote microbial growth.

The use of high temperature may reduce the need for application of mechanical forces, such as turbulence in the wash water. The chemicals commonly used for cleaning are surfactants or alkalis, used to suspend and dissolve the food residues by reducing the surface tension, emulsify fats, and denature proteins [66].

In addition to the mechanical action, other measures must be taken to prevent and control microbial adhesion. In this sense, the facilities, equipment, and utensils should be washed daily and disinfected with the use of microbicides substances previously approved by legislation.

However, there are studies showing that even using the recommended concentration of sanitizing, resistance of bacteria in biofilms still exists. A study performed by Ref. [9] evaluated the bactericidal capacity of peracetic acid on *Salmonella* biofilm and noted that 44.11% (15/34) of the strains were susceptible to the disinfectant in the concentration of 0.2%. In other strains, that were resistant to disinfectants (55.89%, 19/34), more concentrated solutions of the disinfectant were tested, 0.3–0.5%, and for the concentration of 0.5% was observed that was the maximum sanitizing capacity, with no resistant isolate to the agent. This results, generates concern, because when it is used at the indicated concentration by the manufacturer and by the regulatory agency (0.2%), there is inefficiency of the product, and may result in contamination of the surfaces, maintaining the bacteria in the cutting processing environment of chicken.

The disinfection is the use of products for elimination of microorganisms, especially pathogenic. The purpose of disinfection is to reduce the microbial load remaining on the surface after cleaning and prevents their proliferation before restarting the production process. Disinfectants must be effective, safe, and easy to handle, they should be removed from surfaces easily, using water, leaving no residue in the final product that may affect the consumer [68].

Mechanism of action	Examples	Compounds
Blocking in the bacterial adhesion	Policides	2-pyridone bicyclic
	Iron chelators	Lactoferrin; plant extracts, tannins
QS inactivation	Competition for receptor sites/AIs degradation	Furanones halogenated and peptide inhibitor of RNA III (RIP)
Mature Biofilms—Matriz	Enzymes	Proteinase, typsin, DNAase, sodium metaperiodate
	Alteração no pH	Detergentes ácidos/alcalinos
Mature Biofilms- Biomass	Nanoparticles	Zinc, silver, titanium, gold.
	Antiseptics	Chlorhexidine, triclosan.
	Bioactive	AMP, terpinen-4-o1.

Source: [81] with modification.

Table 2. Key targets to combat microbial biofilms and examples of agents.

The chemicals currently used in disinfection processes belong to the following types: acidic compounds, biocides, aldehyde-based, caustics, chlorine, hydrogen peroxide, iodine, isothiazolinones, ozone, peracetic acid, phenols, biguanides, and surfactants [64, 69]. Some examples of agents that may be used to control and/or eliminate biofilms of *Salmonella* and *Campylobacter* are shown in **Table 2**.

The strategies most used in industry involve the removal of biofilms already installed, by removing the matrix and/or bacterial biomass. As a first step is quoted to use hygienic processes with enzymatic detergents and compounds that promote the sudden change in pH and subsequent matrix liquefaction [70].

The use of enzyme-based detergents may be useful to improve the cleaning process. However, due to the heterogeneity in biofilm matrices, it is necessary to know the exact composition for which suitable enzymatic treatments can be applied [71], so that a mixture of different enzymes can increase the spectrum action on biofilm degradation. These enzymatic processes have the advantage of disaggregate biofilm agglomerates, rather than just remove them from the surface, as is the case of mechanical action.

Another important point to be analyzed for the elimination of bacteria in mature biofilm is the involvement of strain-dependent characteristics, since there are molecular intrinsic factors that may act by preventing the effectiveness of the agents, hindering its penetration depending on the composition of the matrix, and also the mechanism of action of the applied agent.

In general, the policides act by inhibiting adhesin and essential fimbriae synthesis in the process of fixing the bacteria to surfaces. The iron chelating agents prevent the availability of this element in the initial process of accession, essential for the biofilm formation. Inactivation of the *quorum-sensing* system involves the use of compounds that compete for binding sites of self-inducing molecules or direct degradation of these molecules.

The surfactants and biosurfactants are also alternatives that can be used in combating biofilm formation. A study of [72] reported that pretreated surfaces with surfactants may have potential higher than 90% in the prevention of bacterial adhesion, and biosurfactants such as rhamnolipids and short chain fatty acids can promote rupture on biofilms [73, 74]. Since surfactin from *Bacillus subtilis* disperses and prevents the formation of biofilms of *Salmonella enterica, E. coli,* and *Proteus mirabilis* [75].

The nanoparticles, as well as the antimicrobial peptides (AMPs), appears as a current strategy for the removal of biomass of biofilms, since they are stable at high temperature and pressures, have inactivates potential, can easily penetrate the matrix, are less likely to develop resistance, have minimal effect on the human cells and can be used to extend the shelf life of fresh and meat products [76, 77].

Combinations of different treatments, with different types of actions are also useful. For example, ultrasound waves [78] were associated with the improvement performance of proteolytic enzymes. These processes target the biofilm matrix, causing the disaggregation and dispersion of the biomass. However, they are not efficient in eliminating these microorganisms, which can adhere to the surface again and restart a new cycle of the biofilm formation.

Alternatively, under increasing interest for biofilm control is to use bacteriophages, which are viruses with high specificity that infect and lyse bacteria and diffuse easily into the matrix layers, including in mature biofilms [79–81]. This technology is still in development, so information about the bacteriophage action in biofilms is still scarce [82]. However, it is known that the infection of biofilm by a phage depends on their chemical composition and also environmental factors such as temperature, growth phase, media, and phage concentration [83].

Studies on the use of natural antimicrobials as antibiofilm substances, for example, compounds extracted from aromatic plants [84], which are recognized as safe for not leave toxic residues for the consumer and does not change the quality of final product. These compounds have demonstrated their antimicrobial activity in planktonic bacteria and some is being evaluated for its potential in eradicating biofilms.

A research performed by Ref. [85] tested the influence of carvacrol, a broad spectrum antimicrobial found in essential oils of herbs such as oregano and thyme, on biofilm of *S. aureus* and *S.* Typhimurium in stainless steel, and found that carvacrol has inhibitory effect on both species, with the effectiveness of the product associated with the species and stage of biofilm formation, the concentration of application, and the form of treatment.

The use of combined actions involving two or more types of chemical, physical, and natural treatments have been reported as the measure of control with more effectiveness against biofilm formation [86]. These treatments can synergistically enhance and broaden the spectrum of actions to eradicate biofilms.

9. Conclusion

Despite several options for new treatments to prevent and remove biofilms, further studies need to be carried out continuously to understand the dynamics of these structures.

Whereas biofilms are constant sources of contamination of production systems for spoilage and pathogens, having economic and public health impacts, prevention should be included in the objectives of the quality of industrial controls. Among the actions required in all strategies, should be included the frequent monitoring, and internal policies to ensure compliance with the preestablished hygiene plans, particularly, respecting the intervals between cleaning processes.

Acknowledgements

To Fundação de Amparo à Pesquisa do Estado de Minas Gerais (FAPEMIG) for the financial support.

Author details

Daise A. Rossi*, Roberta T. Melo, Eliane P. Mendonça and Guilherme P. Monteiro

*Address all correspondence to: daise.rossi@ufu.br

Laboratory of Applied Animal Biotechnology, Faculty of Veterinary Medicine, Federal University of Uberlândia, Uberlândia, Brazil

References

[1] Pometto AL, Demirci A. Biofilms in the food environment. 2nd ed. New Jersey: John Wiley & Sons; 2015. 320 p. DOI: 10.1002/9781118864036

[2] Sutherland IW. The biofilm matrix—an immobilized but dynamic microbial environment. Trends in Microbiology. 2001;9:222–227. DOI: http://dx.doi.org/10.1016/S0966-842X(01)02012-1

[3] Azevedo NF, Cerca N. Biofilms in: health, environment, industry. 1st ed. Porto: Publindústria Technical Issues: 2012. 396 p.

[4] Carpentier B, Cerf O. Biofilms and their consequences, with particular reference to hygiene in the food industry. Journal of Applied Microbiology. 1993;75:499–511. DOI: 10.1111/j.1365-2672.1993.tb01587.x

[5] Jay JM. Food Microbiology. 6th ed. Porto Alegre: Artmed; 2005. 711 p.

[6] Hoiby N, Bjarnsholt T, Givskov M, Molin S, Ciofu O. Antibiotic resistance of bacterial biofilms. International Journal of Antimicrobial Agents. 2010;35:322–332. DOI: http://dx.doi.org/10.1016/j.ijantimicag.2009.12.011

[7] Giaouris E, Chorianopoulos N, Nychas GJE. Effect of temperature, pH, and water activity on biofilm formation by *Salmonella enterica* Enteritidis PT4 on stain less steel surfaces as indicated by the bead vortexing method and conduct ance measurements. Journal of Food Protection. 2012;68:2149–2154.

[8] Steenackers H, Hermans K, Vanderleyden J, Keersmaecker SCJ. *Salmonella* biofilms: An overview on occurrence, structure, regulation and eradication. Food Research International. 2012;45:502–531. DOI: 10.1016/j.foodres.2011.01.038

[9] Vivian RC. The evaluation of biofilm formation and sensitivity to peracetic acid of *Salmonella* spp. isolated from poultry abattoir [dissertation]. São Paulo, Universidade Estadual Paulista, 2014.

[10] Watnick P, Kolter R. Minireview: Biofilm, city of microbes. Journal of Bacteriology. 2000;182:2675–2679. DOI: 10.1128/JB.182.10.2675-2679.2000

[11] Lewis K. Riddle of biofilm resistance. Antimicrobial Agents and Chemotherapy. 2001;45:999–1007. DOI: 10.1128/AAC.45.4.999-1007.2001

[12] Jass J, Walker J. Biofilms and biofouling. In: Walker J, Surmanand S, Jass J, editors. Industrial biofouling: Detection, prevention and control. New Jersey: John Wiley & Sons; 2000. pp.1–12. ISBN: 978-0-471-98866-3

[13] CDC. Foodborne illness surveillance, response, and data systems. Atlanta: U.S. Department of Health and Human Services. 2014. Available in: http://www.cdc.gov/foodborneburden/surveillance-systems.html [Accessed: 2016-06-01]

[14] Baugh S. The role of multidrug efflux pumps in biofilm formation of *Salmonella enterica* serovar Typhimurium. [thesis] Birmingham: University of Birmingham; 2013.

[15] Quinones B, Miller WG, Bates AH, Mandrell RE. Autoinducer-2 production in *Campylobacter jejuni* contributes to chicken colonization. Applied and Environmental Microbiology. 2009;75:281–285. DOI: 10.1128/AEM.01803-08

[16] Petrova OE, Sauer K. Sticky situations: Key components that control bacterial surface attachment. Journal of Bacteriology. 2012;194:2413–2425. DOI: 10.1128/JB.00003-12

[17] Skandamis PN, Nychas GJ. Quorum sensing in the context of food microbiology. Applied and Environmental Microbiology. 2012;78:5473–82. DOI: 10.1128/AEM.00468-12

[18] Srey S, Jahid IK, Ha S. Biofilm formation in food industries: A food safety concern. Food Control. 2013;31:572–585. DOI: 10.1016/j.foodcont.2012.12.001

[19] Malaeb L, Le-Clech P, Vrouwenvelder JS, Ayoub GM, Saikaly PE. Do biological-based strategies hold promise to biofouling control in MBRs? Water Research. 2013;47:5447–5463. DOI: 10.1016/j.watres.2013.06.033

[20] Matyar F, Gülnaz O, Guzeldag G, Mercimek HA, Akturk S, Arkut A, Sumengen M. Antibiotic and heavy metal resistance in Gram-negative bacteria isolated from the Seyhan Dam Lake and Seyhan River in Turkey. Annals of Microbiology. 2014;64:1033–1040. DOI: 10.1007/s13213-013-0740-8

[21] Moe KK, Mimura J, Ohnishi T, Wake T, Yamazaki W, Nakai M, Misawa N. The mode of biofilm formation on smooth surfaces by *Campylobacter jejuni*. Journal of Veterinary Medical Science. 2010;72:411–416. DOI: 10.1292/jvms.09-0339

[22] Haddock G, Mullin M, MacCallum A, Sherry A, Tetley L, Watson E, Dagleish M, Smith DGE, Everest P. *Campylobacter jejuni* 81-176 forms distinct microcolonies on in vitro-infected human small intestinal tissue prior to biofilm formation. Microbiology. 2010;156:3079–3084. DOI: 10.1099/mic.0.039867-0

[23] Vesterlund S, Paltta J, Karp M, Ouwehand AC. 2005. Measurement of bacterial adhesion in vitro evaluation of different methods. Journal of Microbiological Methods. 2005;60:225–233. DOI: 10.1016/j.mimet.2004.09.013

[24] Cappitelli F, Polo A, Villa F. Biofilm formation in food processing environments is still poorly understood and controlled. Food Engineering Reviews. 2014;6:29–42. DOI: 10.1007/s12393-014-9077-8

[25] Sulaeman S, Hernould M, Schaumann A, Coquet L, Bolla JM, De E, Tresse O. Enhanced adhesion of *Campylobacter jejuni* to abiotic surfaces is mediated by membrane proteins in oxygen-enriched conditions. Plos One. 2012;7:e46402. DOI: 10.1371/journal.pone.0046402

[26] Theoret JR, Cooper KK, Zekarias B, Roland KL, Law BF, Curtiss R, Joens LA. (2012) The *Campylobacter jejuni* dps homologue is important for in vitro biofilm formation and cecal colonization of poultry and may serve as a protective antigen for vaccination. Clinical and Vaccine Immunology. 2012;19:1426–1431. DOI: 10.1128/CVI.00151-12

[27] Monds RD, O'Toole GA. The developmental model of microbial biofilms: Ten years of a paradigm up for review. Trends in Microbiology. 2009;17:73–87. DOI: 10.1016/j.tim.2008.11.001

[28] Reuter M, Mallett A, Pearson BM, Van Vliet AHM. Biofilm formation by *Campylobacter jejuni* is increased under aerobic conditions. Applied and Environmental Microbiology. 2010;76:2122–2128. DOI: 10.1128/AEM.01878-09

[29] Gilbert P, Allison DG, Mcbain AJ. Biofilms *in vitro* and *in vivo*: Do singular mechanisms imply cross-resistance? Journal of Applied Microbiology. 2002;92:98S–110S. DOI: 10.1046/j.1365-2672.92.5s1.5.x

[30] Vidal DR, Ragot C, Thibault F. Bacterial biofilms and resistance to disinfectants. Annales Pharmaceutiques Françaises.1997;55:49–54.

[31] Hanning I, Jarquin R, Slavik M. *Campylobacter jejuni* as a secondary colonizer of poultry biofilms. Journal of Applied Microbiology. 2008;105:1199–1208. DOI: 10.1111/j.1365-2672.2008.03853

[32] Teh KH, Flint S, French N. Biofilm formation by *Campylobacter jejuni* in controlled mixed-microbial populations. International Journal of Food Microbiology. 2010;143:118–124. DOI: 10.1016/j.ijfoodmicro.2010.07.037

[33] Culotti A, Packman AI. *Pseudomonas aeruginosa* facilitates *Campylobacter jejuni* growth in biofilms under oxic flow conditions. FEMS Microbiology Ecology. 2015;91:fiv136. DOI: 10.1093/femsec/fiv136

[34] Hanning I. Capture, survival, colonization and virulence of *Campylobacter jejuni* in poultry biofilms [dissertation]. Arkansas: University of Arkansas; 2008

[35] Manning L, Chadd SA, Baines RN. Key health and welfare indicators for broilers production. World's Poultry Science Journal. 2007;63:47–62. DOI: 10.1079/WPS2005126

[36] Pattison M. Practical intervention strategies for Campylobacter. Journal of Applied Microbiology. 2001;90:121–125. DOI: 10.1046/j.1365-2672.2001.01360.x

[37] Sparks NHC. The role of the water supply system in the infection and control of Campylobacter in chicken. World's Poultry Science Journal. 2009;65:459–473. DOI: 10.1017/S0043933909000324

[38] Ogden ID, Macrae M, Johnston M, Strachan NJC, Cody AJ, Dingle KE, Newell DG. Use of multilocus sequence typing to investigate the association between the presence of Campylobacter spp. in broiler drinking water and Campylobacter colonization in broilers. Applied and Environmental Microbiology. 2007;73:5125–5129.

[39] FAO/WHO. 2009. *Salmonella* and *Campylobacter* in chicken meat. Geneva: Microbiological Risk Assessment Series No. 19. 69 pp. Available in: http://www.who.int/foodsafety/publications/micro/MRA19.pdf [Accessed: 2016-05-03]

[40] Jang KI, Kim MG, Ha SD, Lee KA, Chung DH, Kim CH, Kim KY. Morphology and adhesion of *Campylobacter jejuni* to chicken skin under varying conditions. Journal of Microbiology and Biotechnology. 2007;17:202–206.

[41] Habimana O, Møretrø T, Langsrud S, Vestby LK, Nesse LL, Heir E. Micro ecosystems from feed industry surfaces: A survival and biofilm study of Salmonella versus host resident flora strains. BMC Veterinary Research. 2010;2: 48.

[42] Vestby LK, Lönn-Stensrud J, Møretrø T, Langsrud S, AamdalScheie A, Benneche T, Nesse LL. A synthetic furanone potentiates the effect of disinfectants on Salmonella in biofilm. Journal of Applied Microbiology. 2010;108:771–778.

[43] Ewing CP, Andreishcheva E, Guerry P. Functional characterization of flagellin glycosylation in *Campylobacter jejuni* 81-176. Journal of Bacteriology. 2009;191:7086–7093. DOI: 10.1128/JB.00378-09

[44] Rathbun KM, Hall JE, Thompson SA. Cj0596 is a periplasmic peptidyl prolyl cis-trans isomerase involved in *Campylobacter jejuni* motility, invasion and colonization. BMC Microbiology. 2009;9:160. DOI: 10.1186/1471-2180-9-160

[45] Naito M, Frirdich E, Fields JA, Pryjma M, Li J, Cameron A, Gilbert M, Thompson SA, Gaynor EC. Effects of sequential *Campylobacter jejuni* 81-176 lipooligosaccharide core truncations on biofilm formation, stress survival, and pathogenesis. Journal of Bacteriology. 2010;192:2182–2192. DOI: 10.1128/JB.01222-09

[46] Rajashekara G, Drozd M, Gangaiah D, Jeon B, Liu Z, Zhang Q. Functional characterization of the twin-arginine translocation system in *Campylobacter jejuni*. Foodborne Pathogens Diseases. 2009;6:935–945. DOI: 10.1089/fpd.2009.0298

[47] Kalmokoff M, Lanthier P, Tremblay TL, Foss M, Lau PC, Sanders G, Austin J, Kelly J, Szymanski CM. Proteomic analysis of *Campylobacter jejuni* 11168 biofilms reveals a role for the motility complex in biofilm formation. Journal of Bacteriology. 2006;188:4312–4320. DOI: 10.1128/JB.01975-05

[48] Sampathkumar B, Napper S, Carrillo CD, Willson P, Taboada E, Nash JH, Potter AA, Babiuk LA, Allan BJ. Transcriptional and translational expression patterns associated with immobilized growth of *Campylobacter jejuni*. Microbiology. 2006;152:567–577. DOI: 10.1099/mic.0.28405-0

[49] Gibson DL, White AP, Rajotte CM, Kay WW. AgfC and AgfE facilitate extracellular thin aggregative fimbriae synthesis in *Salmonella* Enteritidis. Society for General Microbiology. 2007;153:1131–1140. DOI: 10.1099/mic.0.2006/000935-0

[50] Hamilton S, Bongaerts RJ, Mulholland F, Cochrane B, Porter J, Lucchini S, Lappin-Scott HM, Hinton JCD. The transcriptional programme of *Salmonella enterica* serovar Typhimurium reveals a key role for tryptophan metabolism in biofilms. BMC Genomics. 2009;10:599. DOI: 10.1186/1471-2164-10-599

[51] Barnhart MM, Chapman MR. Curli biogenesis and function. Annual Review of Microbiology. 2006;60:131–147. DOI: 10.1146/annurev.micro.60.080805.142106

[52] Gerstel U, Römling U. The csgD promoter, a control unit for biofilm formation in *Salmonella* Typhimurium. Research in Microbiology. 2003;154:659–667. DOI: 10.1016/j.resmic.2003.08.005

[53] Zogaj X, Nimtz M, Rohde M, Bokranz W, Romling U. The multicellular morphotypes of *Salmonella* Typhimurium and *Escherichia coli* produce cellulose as the second component of the extracellular matrix. Molecular Microbiology. 2001;39:1452–1463. DOI: 10.1046/j.1365-2958.2001.02337.x

[54] Guiney DG. Regulation of bacterial virulence gene expression by the host environment. Journal of Clinical Investigation. 1997;99:565–569.

[55] Reeser RJ, Medler RT, Billington SJ, Jost BH, Joens LA. Characterization of *Campylobacter jejuni* biofilms under defined growth conditions. Applied and Environmental Microbiology. 2007;73:1908–1913. DOI: 10.1128/AEM.00740-06

[56] Svensson SL. Molecular mechanisms of *Campylobacter jejuni* survival: Characterization of the CprRS two-component regulatory system and biofilm formation. Doctorate Thesis, Lakehead University, Thunder Bay, Ontario; 2012. 152 pp.

[57] Ica T, Caner V, Istanbullu O, Nguyen HD, Ahmed B, Call DR, Beyenal H. Characterization of mono- and mixed-culture *Campylobacter jejuni* Biofilms. Applied and Environmental Microbiology. 2012;78:1033–1038. DOI: 10.1128/AEM.07364-11

[58] Donlan RM, Costerton JM. Biofilms: Survival mechanisms of clinically relevant microorganisms. Clinical Microbiology Review. 2002;15:167–193. DOI: 10.1128/CMR.15.2.167-193.2002

[59] Gonzalez JE, Keshavan ND. Messing with bacterial quorum sensing. Microbiology and Molecular Biology Reviews. 2006;70:859–875. DOI: 10.1128/MMBR.00002-06

[60] Gölz G, Adler L, Huehn S, Alter T. LuxS distribution and AI-2 activity in *Campylobacter* spp. Journal of Applied Microbiology. 2012a;112:571–578. DOI: 10.1111/j.1365-2672.2011.05221

[61] Gölz G, Sharbati S, Backert S, Alter T. Quorum sensing dependent phenotypes and their molecular mechanisms in Campylobacterales. European Journal of Microbiology and Immunology. 2012b;2:50–60. DOI: 10.1556/EuJMI.2.2012.1.8

[62] Chmielewski RAN, Frank JF. A predictive model for heat inactivation of *Listeria monocytogenes* biofilm on rubber. LWT – Food Science and Technology. 2006;39:11–19. DOI: 10.1016/j.lwt.2004.10.006

[63] Dosti B, Guzel-Seydim Z, Greene AK. Effectiveness of ozone, heat andchlorine for destroying common food spoilage bacteria in synthetic media and biofilms. International Journal of Dairy Technology. 2005;58:19–24. DOI: 10.1111/j.1471-0307.2005.00176.x

[64] Simões M, Simões LC, Machado I, Pereira MO, Vieira MJ. Control of flow-generated biofilms using surfactants – Evidence of resistance and recovery. Food and Bioproducts Processing. 2006;84:338–345. DOI: 10.1205/fbp06022

[65] Pereira A, Mendes J, Melo LF. Using nanovibrations to monitor biofouling. Biotechnology and Bioengineering. 2008;15:1407–1415. DOI: 10.1002/bit.21696

[66] Maukonen J, Matto J, Wirtanen G, Raaska L, Mattila-Sandholm T, Saarela M. Methodologies for the characterization of microbes in industrial environments: A review.

Journal of Industrial Microbiology and Biotechnology. 2003;30:327–356. DOI: 10.1007/s10295-003-0056-y

[67] Forsythe SJ, Hayes PR. Food hygiene, microbiology and HACCP. 3rd ed. Nottingham: Aspen Publishers; 1998. 17 p. DOI: 10.1007/978-1-4615-2193-8

[68] Simoes LC, Simoes M, Vieira MJ. Influence of the diversity of bacterial isolates from drinking water on resistance of biofilms to disinfection. Applied Environmental Microbiology. 2010;76:6673–6679. DOI: 10.1128/AEM.00872-10

[69] Bremer PJ, Fillery S, McQuillan AJ. Laboratory scale clean-in-place (CIP) studies on the effectiveness of different caustic and acid wash steps on the removal of dairy biofilms. International Journal of Food Microbiology. 2006;106:254–262. DOI: 10.1016/j.ijfoodmicro.2005.07.004

[70] Machado SMO. Evaluation of the antimicrobial effect of benzalkonium chloride surfactant in controlling the formation of undesirable biofilms. [dissertation] Minho: Universidade do Minho, 2005.

[71] Bridier A, Briandet R, Thomas V, Dubois-Brissonnet F. Resistance of bacterial biofilms to disinfectants: A review. Biofouling The Journal of Bioadhesion and Biofilm Research. 2011;27:1017–1032. DOI: 10.1080/08927014.2011.626899

[72] Cloete TE, Jacobs L. Surfactants and the attachment of Pseudomonas aeruginosa to 3CR12 stainless steel and glass. Water SA, 2001;27:21–26. ISSN: 0378-4738

[73] Davies DG, Marques CNH. A fatty acid messenger is responsible for inducing dispersion in microbial biofilms. Journal of Bacteriology. 2009;191:1393–1403. DOI: 10.1128/JB.01214-08

[74] Dusane DH, Nancharaiah YV, Zinjarde SS, Venugopalan VP. Rhamnolipid mediated disruption of marine Bacillus pumilus biofilms. Colloids and Surfaces B: Biointerfaces. 2010;81:242–248. DOI: 10.1016/j.colsurfb.2010.07.013

[75] Mireles JR, Toguchi A, Harshey RM. *Salmonella* enterica serovar Typhimurium swarming mutants with altered biofilm-forming abilities: Surfactin inhibits biofilm formation. Journal of Bacteriology. 2001;183:5848–5854. DOI: 10.1128/JB.183.20.5848-5854.2001

[76] Costa F, Carvalho IF, Montelaro RC, Gomes P, Martins MCL. Covalent immobilization of antimicrobial peptides (AMPs) onto biomaterial surfaces. Acta Biomaterialia. 2011;7:1431–1440. DOI: 10.1016/j.actbio.2010.11.005

[77] Lu X, Samuelson DR, Rasco BA, Konkel ME. Elucidation of the antimicrobial effect of diallyl sulfide on *Campylobacter jejuni* biofilms. Journal of Antimicrobial Chemotherapy. 2012;67:1915–1926. DOI: 10.1093/jac/DKS138

[78] Oulahal-Lagsir O, Martial-Gros A, Bonneau M, Blum LJ. "Escherichia coli-milk" biofilm removal from stainless steel surfaces: Synergism between ultrasonic waves and enzymes. Biofouling. 2003;19:159–168. DOI: 10.1080/08927014.2003.10382978

[79] Kudva I T, Jelacic S, Tarr PI, Youderian P, Hovde CJ. Biocontrol of *Escherichia coli* O157 with O157-specific bacteriophages. Applied and Environmental Microbiology, 1999;65:3767–3773.

[80] Briandet R, Lacroix-Gueu P, RenaultM, Lecart S, Meylheuc T, Bidnenko E, Steenkeste K, Bellon-Fontaine MN, Fontaine-Aupart MP. Fluorescence correlation spectroscopy to study diffusion and reaction of bacteriophages inside biofilms. Applied and Environmental Microbiology. 2008;74:2135–2143. DOI: 10.1128/AEM.02304-07

[81] Donlan RM. Preventing biofilms of clinically relevant organisms using bacteriophage. Trends in Microbiology. 2009;17:66–72. DOI: 10.1016/j.tim.2008.11.002

[82] Sutherland IW, Hughes KA, Skillman LC, Tait K. The interaction of phage and biofilms. FEMS Microbiology Letters. 2004;232:1–6. DOI: 10.1016/S0378-1097(04)00041-2

[83] Chaignon P, Sadovskaya I, Ragunah CH, Ramasubbu N, Kaplan JB, Jabbouri S. Susceptibility of staphylococcal biofilms to enzymatic treatments depends on their chemical composition. Applied Microbiology and Biotechnology. 2007;75:125–132. DOI: 10.1007/s00253-006-0790-y

[84] Lewis K, Ausubel FM. Prospects for plant-derived antibacterials. Nature Biotechnoloy. 2006;24:1504–1507. DOI: 10.1038/nbt1206-1504

[85] Knowles JR, Roller S, Murray DB, Naidu AS. (2005). Antimicrobial action of carvacrol at different stages of dual-species biofilm development by Staphylococcus aureus and *Salmonella enterica* serovar Typhimurium. Applied and Environmental Microbiology. 2005;71:797–803. DOI: 10.1128/AEM.71.2.797-803.2005

[86] Nazer A, Kobilinsky A, Tholozan JL, Dubois-Brissonnet F. Combinations of food antimicrobials at low levels to inhibit the growth of Salmonella sv.Typhimurium: A synergistic effect? Food Microbiology. 2005;22:391–398. DOI: 10.1016/j.fm.2004.10.003

Selenium Requirements and Metabolism in Poultry

Anicke Brandt-Kjelsen, Brit Salbu, Anna Haug and
Joanna Szpunar

Abstract

As counteract against deficiency in livestock, dietary treatments are supplemented with selenium (Se), usually as the inorganic form sodium selenite (SS). Since Se is considered as toxic as well as an essential element, SS is added to prevent an increase of Se in edible chicken parts. However, in many countries, populations suffer from suboptimal Se intake and even Se deficiency, by increasing the use of organic Se sources such as Se-enriched yeast or wheat in animal feed, there will be a subsequent increase in meat and egg products for human consumption. One could argue that the chickens do not need the extra pool of Se in muscles, as the inorganic form will be sufficient to meet the chickens Se requirements. Since the feed is fortified with selenite, the chickens will always have adequate access to the essential trace element. However, global gene set functional enrichment analysis revealed statistically significant enrichment of a number of biological processes that were dependent on the Se feed sources, such as cell growth, organ development and protein metabolism in favour of organic Se.

Keywords: selenium, requirements, metabolism, feed supplementation, selenoamino acids

1. Introduction

1.1. The element selenium

Selenium (Se) was discovered in 1817 by the Swedish chemist Jöns Jakob Berzelius [1, 2]. The element was named after the moon in Greek; Selene. Selenium is the chemical element number 34, a metalloid with six stable isotopes 76, 77, 78, 79, 80 and 82. The element is in the 16 group in the periodic table together with sulphur (S), oxygen (O) and tellurium (Te). These elements have many common chemical properties. Selenium and S have similar electronegativities, atomic radius and the same oxidation states (-2, 0, +4, +6). Selenium is an S analogue, and substitute S in a series of compounds, such as sulphate (selenate), sulphite (selenite), sulphenic acid (selenic acid). Selenium and S react easily with each other forming selenylsulphide bounds [1]. Despite these similar properties, there are some major differences between S and Se, and substitution of one another results in different chemical properties. One of these differences is illustrated by the acid dissociation constant (pKa) value; H_2Se has a pKa of 3.73 and is a much stronger acid than H_2S with a pKa of 6.96 [1]. Another difference between the two elements is the reduction potential, under standard conditions Se is more readily reduced than S (e.g. SeO_4^{2-} has an $E =$ 1.15 volt and SO_4^{2-} has $E = 0.158$ V).

1.2. Seleno amino acids

Cysteine (Cys) and methionine (Met) are S amino acids and both have Se analogs, SeCys and SeMet, respectively. As for the acids, the amino acids exhibit different chemical properties depending on the properties of Se and S. While the chemical properties of SeMet and Met are quite similar [3] the pKa (amino group) of SeCys is much lower than for Cys (5.2 vs. 8.3) [4]. At physiological pH the selenol group of SeCys will be in its anionic selenolate form, while the thiols of Cys residues are protonated. These properties make SeCys a more reactive amino acid compared to Cys [1], and give unique Se derived properties to selenoproteins.

1.3. Selenium in the terrestrial environment

High Se concentrations are found in sedimentary rocks and shales formed dung the cretaceous period, while lower concentrations are characteristic for igneous (volcanic) rock, sandstone, granite and lime stone [5, 6]. The ubiquitous but uneven Se distribution in bedrock has resulted in Se concentrations in soil ranging from almost zero to 1250 mg Se kg^{-1} in some seleniferous soils in Ireland [7, 8]. High soil Se concentrations can also be a consequence of prolonged deposition from precipitation containing seawater. I Norway the observed Se concentration in humus showed that the Se concentrations in soil decreases with distance from the sea [9]. Through observations in animal production, it is apparent that vast land areas worldwide do not supply sufficient Se for optimal livestock nutrition, and areas with low Se concentrations in soil are much more common than areas with very high Se concentrations in soil [8].

1.3.1. Selenium transfer from soil to plant

The distribution and availability of Se to plants depend on different soil conditions such as pH, organic matter [10], redox conditions, competing ionic species (e.g. sulphate), microbial activity, soil texture, compaction, mineralogy, temperature and moisture [11]. The bioavailability of Se from soil is more important than the total soil Se concentration. Inorganic Se species present in soil are selenide (Se_2^-), elemental Se (Se_0), selenate (SeO_4^{2-}) and selenite (SeO_3^{2-}) [12]. In aerobic soils, with pH around seven, selenate predominates, whereas selenite predominates at lower pH and redox potential. Under strongly reduced soil conditions selenide dominates [13]. Selenate is more mobile and thus available to plants compared to selenite.

1.3.2. Selenium in plants

The uptake of Se in higher plants depends on plant species and their physiological phase of development. Selenate is more accessible to plants than selenite and organic Se such as SeMet. The uptake of selenate and SeMet follow the active transporters of analog S species and is metabolized through the sulphate assimilation pathway. Selenite, however, is not mediated by membrane sulphite transporters [2]. After selenite has been taken up in the plant it is rapidly transformed to SeMet or SeCys and accumulated in proteins in the roots, while selenate is highly mobile in the xylem and translocated to plant parts above ground [12, 14]. The Se concentrations in plants are related to the general protein content of the plants and their different ability to accumulate Se. Most plants that are used in forage as well as grasses are non-accumulators [2] and, hence, have low Se concentrations. Surveys on Se concentrations in crop have revealed that areas producing crops with Se concentrations too low (<0.1 mg Se kg^{-1}) to meet animal requirements are more common than areas producing toxic levels (>2 mg Se kg^{-1}) of Se in crops, hence supplementation of diets is necessary to meet the animals Se requirements. There has been an increased interest of investigating the beneficial use of medical plants and phytogenetic compunds in poultry diets [15], some of these plants are Se hyperaccumulators [16] and future research will prove if these plants could replace the traditional supplementation used today.

1.3.3. Selenium requirements in poultry and domestic animals

The toxicity of Se in animal diets was recognized around 1932 [17]. However, a decade later it was obvious that Se deficiency lead to several negative effect. Impaired growth and development, poor feathering, reduced egg production and hatchability, pancreatic degeneration, nutritional muscular dystrophy, and necrotic lesions in liver, muscle and heart are among the effects reported in Se deficient animals, all of which resulted in large economic losses for the industry [18–20]. In Finland, inadequate Se intake caused nutritional disorders in pigs and reduced the profitability of the production. As a result, all commercial animal feeds have been supplemented with selenite since 1969 [8], followed by Denmark in 1975, Norway in 1979 [21] and Sweden in 1980.

In addition, low Se levels reduce the immune function and the animals are more prone to develop diseases such as white muscle disease, cardiac muscle metamorphism, blood capillary disease, cancer, anaemia and liver bleeding [22, 23]. In low Se areas worldwide, livestock producers have adopted methods that ensure adequate Se status in animals. Since selenium was foreseen as a toxic element, the supplementation recommendations were designed to ensure that the Se requirements were met without increasing the Se concentrations in tissues significantly. In that way humans were protected from ingesting "toxic levels" of Se from food products [24]. Sodium selenite (SS) and sodium selenate which do not increase muscle Se levels substantially were therefore approved as supplements to the livestock and poultry industry [24]. Selenite is most often used [24]. The use of Se-enriched yeast (SY) as feed additive is increasing. The European Food Safety Authority has approved a maximum of 0.2 mg Se from different yeast strings in complete Feed [25, 26]. Se-enriched yeast approved as supplement consists of 60–85% SeMet, 2–4% SeCys and <1% selenite or selenate [27]. Based on the legislation from the European Committee the allowed maximum total feed Se concentration is 0.5 mg kg⁻¹ DM (dry weight) [26]. These regulations also distinguish between inorganic or organic Se, where the maximum of organic Se is set to 0.2 mg Se kg⁻¹. Numerous reports on the positive effects of organic Se sources in animal feed are available. In addition to reduced frequency of diseases related to Se deficiency, organic Se readily increases tissue Se concentrations [28, 29]. In 2014 DL-Selenomethionine was accepted as an organic Se source after it was showed to be safe for chickens for fattening up to 1.5 mg Se/kg feed, the upper limit of 0.2 mg organic Se/kg feed still stands [30].

2. Understanding the speciation and metabolism of selenium in poultry

2.1. Bioavailability and bioaccessibility of selenium in a sample preparation and instrument analysis perspective

2.1.1. Speciation analysis of selenium

Speciation analysis of Se is applied to identify and quantify Se species present in different matrices. This is of major importance as bioavailability, cross membrane transport and metabolism of Se is highly dependent on the Se species present [31]. While numerous studies on Se species in yeast and Se accumulating plants have been published (e.g. [32–34]), fewer studies have been carried out on Se species in common food. It has therefore been necessary to develop reliable analytical techniques to study the speciation of Se in environmental and biological samples, to understand the biochemical cycle of Se [35, 36].

Selenium is usually present in food and biological samples at low concentrations and can be present as a variety of Se species. To perform speciation analysis of Se in biological matrixes different methodologies are applied. Hyphenated techniques such as liquid chromatography coupled to inductive coupled plasma mass spectrometry (LC-ICP-MS) and liquid chromatography coupled electrospray ionization mass spectrometry (LC-ESI-MS) are two complementary detection methods [37]. The ICP-MS quantifies based on signals from the elements mass/charge ratio; the ionization is virtually species independent and accurate absolute quantification is possible. In ESI-MS the Se species are identified based on molecules mass/charge ratio, where ionization is species and matrix dependent and quantification requires isotopically labelled species. Identification based on retention time matching with well-defined standards by HPLC-ICP MS is tentative, while ESI-MS data alone does not provide sufficient evidence for structural confirmation [33, 38].

With increased knowledge and sensitive instruments, more information on Se species in food will emerge. As many detectors are very sensitive (low detection limits), the challenges lie within sample preparation and chromatographic separations. At present, no standardized approaches to Se speciation analysis exist and different strategies of sample preparation and separation prior to detection are applied.

2.1.2. Sample preparation and separation techniques

Selenium species in food and biological samples can be extracted using different agents. Water extraction has been applied to water-soluble Se amino acids (MeSeCys and γ-glutamyl-MeSeCys), driselase is used to release Se bound to cell walls, different proteases for hydrolysis of selenoproteins and seleno-containing proteins [34], and tryptic digestion of water-soluble Se species are applied to identify peptide sequences [39]. Another approach is the use of in vitro digestion steps mimicking the gastro intestinal tract to get information on the bioaccessible fraction of Se from different matricies [40, 41].

Following extraction, different techniques are used to separate Se species. The most used technique is high performance liquid chromatography (HPLC) with reversed phase (RP), ion-paring (IP)-RP, or ion-exchange (IE) columns, sometimes with a pre-concentration step using a size exclusion column (SEC) to collect the low molecular mass fraction [33, 42, 43]. Optimization of the chromatographic separation conditions to obtain narrow and well defined peaks are strived for at all times for best results [37]. Isotope dilution has improved the quality control of analysis and can be used species-specific or species unspecific, and be applied pre or post column to the ICP-MS [44].

During the last years, measurement of Se has focused on a more proteomic approach revealing information on selenoproteins and seleno-containing proteins. Examples are proteins being separated on gel electrophoresis (GE) (sodium dodecyl sulphate polyacrylamide gel electrophoresis, SDS PAGE) for protein identification by matrix-assisted laser desorption/ionization (MALDI)—time of flight (TOF)-MS or ESI-MS or laser ablation ICP-MS to identify the proteins containing Se in a gel [45]. Another way to identify Se species in a SDS PAGE is the use of radioactive labelled ^{75}Se. Neutron activation analysis (NAA) followed by gamma measurements of Se could give information on trace levels of Se in different matrixes. The use of affinity columns separating selenoalbumin (SelA), selenoprotein P (SelP) and glutathione peroxidase (Gpx) has been successfully reported [45].

Isotopic labelling of Se species makes it possible to follow the fate of different species in the environment, either applied in fertilizers to plants and vegetables, in feed or injected to animals. Detailed knowledge on the formation of different Se species has been gained by this approach [46–50], but some questions remain unanswered such as the turnover of Se form plant crops in adequately fed animals, or the uptake and transformation of Se from colon [36]. Speciation analysis of Se is essential to understand the uptake, metabolism, distribution and transformation of Se in biological system.

2.2. Selenium species dependent metabolism of selenium

According to [51] all Se-forms (chemically pure forms and different forms in food) are generally well transported over the intestinal membrane (70–95%), but the uptake varies according to the Se source and status of the individual [51]. The same active transporters take up selenate as sulphate, while selenite is taken up by passive transport and does not share the pathway of sulphite. Immediately after selenite has entered the red blood cells, it is reduced to selenide by cellular glutathione [52] or through the actions of thioredoxin reductase [53]. Thereafter, selenide is transported to an organ, usually the liver, to undergo selenosynthesis [52]. Selenate reaches the liver in its intact chemical form where it is reduced to selenide through the same pathways as selenite [54]. In the liver of animals inorganic Se and organic Se is transformed into SeCys.

Organic seleno amino acids also reach the liver in their intact form where the conversion to selenide may follow different pathways. SeCys is transformed to selenide through the ν-lyase reaction, while SeMet can follow three known reaction pathways;

1. Conversion to SeCys through the intermediate SeCystathionine.

2. Directly by ν-lyase reaction.

3. SeMet does not enter the selenide pool, but is inserted into proteins as Met by tRNA coding for Met [50, 52, 53]. When SeMet substitutes for Met in proteins the Se concentration increases in the form of Se containing proteins.

Selenide is assumed to be the main precursor of Se metabolism. Selenide has three main conversion routes dependent on the Se status of the organism. Firstly, selenide is transformed to Se-phosphate then to Se-cysteyl tRNA for insertion of SeCys in selenoproteins. Secondly, at lower levels of intake, excess selenide not used in selenoprotein synthesis is also converted into selenosugars for excretion via urine. Thirdly, at high levels of intake, methyltransferases add a methyl group to selenide, leading to a sequential conversion to methylselenol and further to dimethyl selenide (excreted via breath and feces) and trimethyl selenonium (excreted via urine) in addition to excretion of selenosugars [50, 53, 55].

Methylated Se species, MeSeCys, ν-glut-MeSeCys and selenobetaine follow different pathways compared to the other dietary Se species described. The methylated selenospecies are transformed directly into methylselenol and excretion, or are demethylated to selenide [52].

Bioavailability studies should also address the transformation of Se into biological active metabolites (bioactive Se) [51, 52]. Selenoproteins are considered bioactive as they are essential for animals and humans.

Selenoprotein P (SelP) is the major selenoprotein synthesized in the liver and is released to the blood stream, as a transporter of Se in the body. The liver also releases cellular Gpx to the blood stream whereas the kidneys release extracellular Gpx. According to [53] the uptake of SelP from plasma is by specific receptor-mediated processes of apolipoproteins in brain and testis and megalin in kidneys. The specific uptake mechanisms to other tissues remain unknown.

The use of Se supplements to poultry improves the Se nutritional status by increasing the glutathione peroxidase Gpx activity in plasma. Glutathione peroxidase is the commonly used biomarker for Se status in livestock. When birds are fed with selenium enriched yeast the Gpx3 activity in blood remains higher for a longer period of time after supplementation than if the birds are supplemented with sodium selenite [56]. The Se supplementation and Gpx activity in poultry blood follow a dose response relationship [56–58], that seem to level off at approximately 400 ng/g whole blood [59].

2.3. Se supplementation of life stock

As mentioned earlier Se-enriched yeast increase chicken muscle Se concentrations significantly compared with sodium selenite [7, 57]. Hence, increasing the use of organic Se sources such as Se-enriched yeast in animal feed will give a subsequent increase in meat, egg and milk products for human consumption.

The aim of the present work was to compare the ability of Se-enriched wheat (SW), Se-enriched yeast (SY) and sodium selenite (SS), as Se supplements in dietary treatments to increase the Se concentration in edible chicken parts for human consumption. To compare chickens fed elevated Se dietary treatments with industrial produced chickens in Norway. The work also included identification and quantification of Se amino acids (SeMet, SeCys) in the edible parts of the chickens and effects on gene expression.

3. Animal experiment

The experimental research on animals followed internationally recognized guidelines. All animals were cared for according to laws and regulations controlling Norwegian experiments with live animals according to the Norwegian Animal Research Authority.

Ninety male chicken (*Gallus gallus*) (Ross 308, Samvirkelaget kylling, Norway) were divided into three groups and fed with either inorganic Se as sodium selenite (SS) or organic Se as Se-enriched yeast (SY) or Se-enriched wheat (SW). The calculated Se concentration in the dietary treatments was 0.8 mg Se kg^{-1}. Other main dietary constituents were added in equal amounts. Selenite and Se-enriched yeast were used as control groups to the Se-enriched wheat group, as these two Se supplements have been intensively studied at different concentrations levels [58, 60] among others. As a reference five chickens, 28 days of age, and chicken feed were collected from the chicken industry (Nortura SA chicken slaughterhouse in Rakkestad (Østfold, Norway; Ross 308). The industry use SS as the dietary source of Se, the maximum allowed concentration in the diet is 0.5 mg Se kg^{-1} [27].

The chickens were fed ad-libitum for 33 days at the Animal Production Experimental Centre (SHF) at the Norwegian University of Life Sciences (UMB, Ås, Norway) randomly placed in separate pens. The animals were weighted after 8, 11 and 28 days. After 33 days the chickens were slaughtered, muscles (leg and breast) and liver sampled cut out, vacuum packed and stored frozen (-20°C) prior to analysis. Blood samples were taken by beheading and subsequent collection from the neck in 50 ml tubes (average volume of 30 ml), and stored at -20°C. The tissues were freeze dried prior to total Se measurements and speciation analysis.

3.1 Measurements

3.1.1 Total Se measurements

Tissue of muscles (breast and leg) and liver from chickens fed SS, SY or SW were freeze dried and homogenized with a mixer mill (Retsch mixer mill MM 200) equipped with zirconium jars. The samples were digested using a microwave assisted (UltraCLAVE, Milestone) nitric acid (5 ml distilled suprapure 14 M HNO_3) decomposition of 0.1 g of dried tissue, or 1 ml (weight) of whole blood at 240°C for 40 min and diluted to 50 ml with 2% ethanol. Total Se was measured using high resolution inductive coupled plasma mass spectrometry (HR-ICP-MS) (Thermo Finnigan Element2, Bremen, Germany) at the National Institute of Occupational Health (STAMI). The instrument was used in high resolution mode. Tellurium was added as internal standard (IS).

The total concentration in the supernatants were digested (1 ml, weight) with sub-distilled nitric acid using the UltraCLAVE and diluted to 50 ml with MQ water. The samples were mixed 1:1 with 4% ethanol using a mixing block connected to the peristaltic pump of the ICP-MS (Perkin Elmer, Elan 6000) and total Se was measured on mass 82. Wheat flour (1567A) and Bovine liver (1577B) from NIST were used as certified reference material. Tellurium was added as internal standard (IS).

3.1.2 Speciation analysis

Proteolytic extractions were carried out according to the method from [61] on freeze dried and homogenized samples from three different chickens of each dietary treatment. One ml of 0.1 M Tris-HCl (pH 7.5) was added to 0.1 g meat sample and the suspension mixed with an ultrasonic probe for 1 min. Reduction and carboxymethylation (CAM) of the samples was done by adding 200 µl 0.2 M DTT (in 0.1 M Tris-HCl pH 7.5) and 275 µl 0.5 M IAM. The samples were incubated with careful mixing in the dark at room temperature (20°C). After 2 h 2.5 ml 0.2 M DTT was added and the samples shaken for 1 h in order to destroy excess of IAM. Then, 7 ml of 0.1 M Tris-HCl buffer was added together with protease XIV and lipase. The samples were digested with 30 mg protease and 20 mg lipase over night (×3) at 37 ± 0.5°C in an incubator cupboard on a Roto-Shake (Genie) in the dark. All digested samples were centrifuged at 10,000 × g for 10 min at 4°C and supernatants pooled and freeze dried. The freeze dried supernatants were dissolved in 5 ml of MQ water. One ml was used for total Se measurements and 4 ml for identification and quantification of Se amino acids. Enzymatic extracts were centrifuged at 12,110 g for 5 min (mini Spin, Eppendorf) and filtered through a syringe filter (0.45 µm) before high performance liquid chromatography (HPLC, Perkin Elmer 200 series pump)-ICP-MS (Elan 6000) analysis. The chromatographic separation was done on a reversed phase (RP) Altima C8/Alltech (150 × 4.6 mm, 5 µm) column using an injection volume of 100 µl. Isocratic elution of 2% MeOH and 0.1% HFBA (as an ion pairing agent) with a flow rate of 0.8 or 0.9 ml min^{-1}. Separation of CAM SeCys and CAM SeMet were preformed within 10 min.

The SeCys standard was synthesizes according to the procedure described by [61]. Wheat flour (BC210a) from European Reference Material (ERM) from LGC-standards is verified for SeMet and was added to the speciation analysis for quality assurance.

3.1.3 Amino acid measurements

Total (peptide bound and free) amino acids were measured in two breast muscle samples from each treatment (SS, SY and SeW). The measurements were performed on a Biochrom 30 Amino

Acid Analyzer (Biochrom Ltd., Cambridge, UK) at Aquaculture Protein Centre (APC) at NMBU, following the procedure described by the Commission Directive 98/64/EC (1998). The samples were oxidized and hydrolyzed prior to HPLC-ultra violet (UV) measurements. The method does not distinguish between salts of amino acids and cannot differentiate between D and L forms of amino acids.

3.1.4 Gene expression analysis

Gene expression analyses offer a sensitive and rapid detection of transcriptional changes occurring at the cellular level after different treatments and exposures. Global (un-biased) transcriptional analysis using a oligo-array was applied to see whether SS and SW dietary treatments had different effect on the gene expression profiles of chicken muscle in regard to chicken health. The analysis was performed at Norwegian Institute of Water Research (NIVA).

Isolation of mRNA from fresh frozen muscle samples were performed by TRIzol® extraction, quality controlled by measuring salt and phenol interferences and approximately mRNA concentration by nanodrop (Nanodrop® ND-100 UV-Visible spectrophotometer, NanoDrop Technologies). The samples were diluted to 100 ng µL^{-1} so that RNA integrity could be determined by gel electrophoreses (Bioanalyzer Instrument, Agilent Technologies) samples passing the quality cut-off criteria had RNA Integrity Number (RIN) >9, 260/230 > 1.99 and 260/280 > 2.31.

Microarray analysis was performed following Agilent's protocol "One Color Microarray-Based Gene Expression Analysis" (Quick Amp Labelling, Version 5.7 March 2008) using a 44k Agilent Chicken (V2) array.

3.1.5 Statistics

Statistical analysis was conducted using one way ANOVA with 95% confidence interval and Tukeys simultaneous test to establish statistical difference ($p < 0.05$) among individual treatment means using GraphPad prism version 5.

3.2 Results and discussion

3.2.1 Feed Se concentrations

The Se concentration in the reference diet from Nortura SA ($n = 6$) was 0.47 ± 0.052 mg Se kg^{-1}. The Se concentrations in the elevated dietary treatments ($n = 6$) were 0.90 ± 0.030 mg Se kg^{-1} for SS, 0.91 ± 0.040 mg Se kg^{-1} for SY, and 1.0 ± 0.017 mg Se kg^{-1} for SW. The Se concentrations were 0.1–0.2 mg Se kg^{-1} higher than the estimated concentration of 0.8 mg Se kg^{-1}, likely attributed to other constituents in the diet such as fishmeal and soy meal.

3.2.2 Bird physiological performance

The mortality was within the expected range (10% in the group fed with SS, 17% in the group fed with SY, and 3% in the group fed with SW) of experiments performed at the Animal

Production Experimental Centre. There were no effects of dietary treatment on intake or live weight gain.

3.2.3 Bioavailability of Se: total Se concentrations

The limit of detection was 0.4 µg L^{-1} based on three times the standard deviation of eight blank samples and the mean standard deviations of Se-77, Se-78 and Se-82. The limit of quantification was based on 10 times the standard deviation of the blank samples and was 1.3 µg L^{-1}.

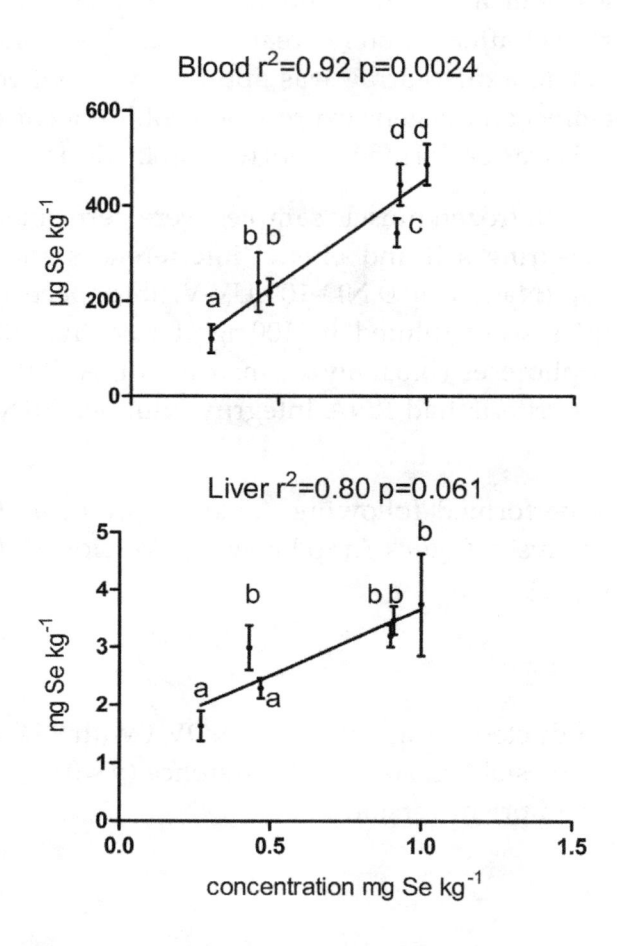

Figure 1. Correlation between feed Se concentration and liver and blood Se concentrations.

Whole blood (wet weight) Se-concentrations were significantly higher ($p < 0.05$) for all the chickens receiving Se elevated dietary treatments (SS, SY and SW) compared to the reference group (reference SS) (**Figure 1**). The Se concentration in whole blood from the chicken receiving SY and SW were 468 ± 43 µg Se kg^{-1} and 521 ± 43 µg Se kg^{-1}, respectively and significantly higher ($p < 0.05$) than the concentration of 362 ± 32 µg Se kg^{-1} in whole blood of the SS group. There were no differences in blood Se-concentrations between the two groups fed with organic Se. The dose dependent increase of Se in blood (**Figure 1**) with increasing Se in the diet is in line with previous published results in chickens and young turkeys [7, 57, 58]. The results are, however, not in line with results from [58, 59] were no difference in whole blood Se concentrations were observed between SS and SY at the same dietary Se level (in lamb and chickens, respectively). The Se concentration in whole blood (220 ± 27 µg Se kg^{-1}) of the reference group was in accordance with the concentration in whole blood (240 µg L^{-1}) of the SS group of [58] with a dietary treatment of 0.42 mg Se kg^{-1}.

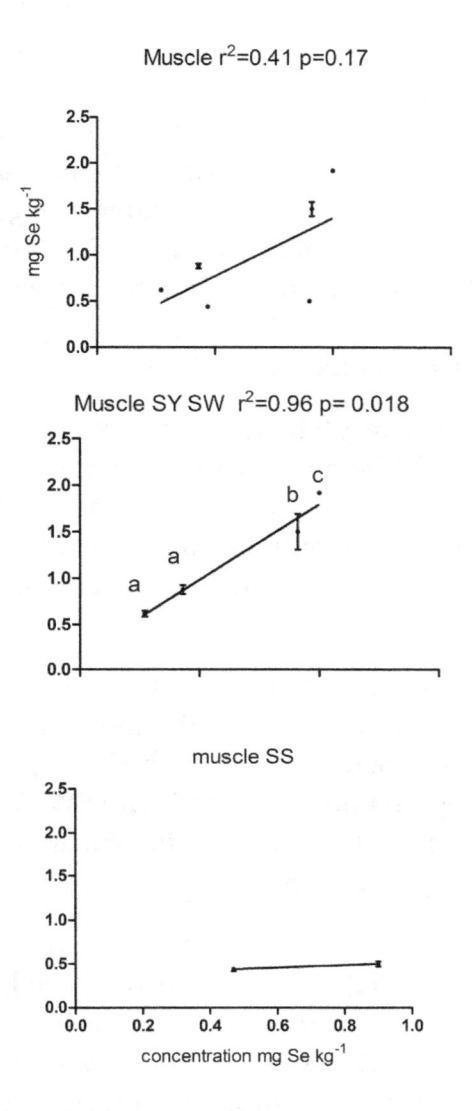

Figure 2. Correlation between Se concentration in feed and Se concentration in muscle with regard to Se source added to the feed (SY, Se-enriched yeast; SW, Se-enriched wheat and SS, sodium senenite).

The total Se concentration differed among tissue types; the liver had the highest Se concentrations, followed by the breast muscle and leg muscle (**Figure 2**). The high liver Se concentration is due to the tendency of glandular visceral tissue to have higher Se concentration than skeletal tissue [59], and the fact that the liver handles most of the absorption of nutrients and regulates their release into blood for further distribution or excretion. The Se concentrations in the livers of the reference group were 2.3 ± 0.18 mg Se kg^{-1}. By increasing the Se concentrations in the diets to 0.9–0.10 mg Se kg^{-1} the Se concentrations in livers increased significantly ($p < 0.05$) compared to the reference group (SS: 3.2 ± 0.19 mg Se kg^{-1}, SY: 3.5 ± 0.25 mg Se kg^{-1}, SW: 3.8 ± 0.89 mg Se kg^{-1}) (**Figure 1**). The equal Se concentrations measured in liver of chickens receiving dietary treatments with different Se forms at the same Se concentrations level was in accordance with results obtained by [49].

An increase from 0.47 ± 0.052–0.9 ± 0.030 mg Se kg^{-1} of Se as SS in the diets did not result in a significant increase in the breast or leg muscle Se concentrations, 0.44 ± 0.24 and 0.50 ± 0.31 mg Se kg^{-1}, respectively (**Figure 2**). Se-concentration in chickens muscle of chickens fed with organic Se was significantly higher ($p < 0.05$) than in muscles from chickens fed with selenite. The Se concentration in breast muscles were significantly higher ($p < 0.05$) in the group receiving SW compared to SY (1.9 ± 0.020 and 1.5 ± 0.019 mg Se kg^{-1}). No difference between SW and SY diets on the total Se concentrations measured in leg muscle. Most publications on Se in dietary treatments use 0.3 and 0.5 mg Se kg^{-1} (e.g. [57]). The muscle Se concentrations in the chickens receiving SS in the diet was in line with the concentrations reported by [62] after dietary supplement with 0.6 mg Se kg^{-1} as SS and 0.6 mg Se kg^{-1} as SY after 42 days, and in line with the total Se concentrations reported by [63] in breast muscle of Ross 308 chickens.

These data show that Se from SW result in similar Se concentrations in muscle as do SY, and based on these criteria SW could be an Se source for biofortification of Se in chickens, as suggested by [64]. The increased concentration of Se in muscle from organic Se in dietary treatments has been shown by several authors, and is attributed to the unspecific uptake of SeMet in the methionine pathway resulting in an unspecific storage of Se in muscle proteins [7, 64–66].

A wet weight portion of 170 g of chicken meat from the industrially produced chickens or selenite supplemented groups would give a contribution of 18 ± 1.3 µg Se, which is close to the values given by the Norwegian Food Table of 13 µg (170 g portion). The contribution from 170 g of chicken meat from chickens supplemented with SW or SY would be 67 ± 7 µg Se, and cover the new Se recommendations from the Nordic Nutrition Recommendations [67] of 50 µg Se/day for women and 60 µg Se/day for men.

The bioactive Se is defined as the fraction of Se from feed or food that is converted into biological active selenoproteins [52]. SeCys is incorporated into all selenoproteins and quantification of SeCys will therefore give a measure of the bioactive amounts of Se in different tissues. Hence, the reservoir of SeMet is not related to a greater bioactivity because of the unspecific incorporation into general body proteins were Se is not recognized by the organism for selenoprotein synthesis [52].

The extraction recovery after enzymatic digestion for the determination of Se-species in muscle was between 94 and 109%, whereas the extraction from liver was lower and varied from 77 to 103%, in feed the extraction was between 83 and 104%. The lower extraction efficiencies in some of the liver and feed samples may be related to a high fat content. The problems with the SeMet derivatization is likely the explanation for the low column recovery (~80%) of total Se compared to the extraction efficiency (>95%) in some of the samples, The different chemical structure of the Se binding to − H in the SeCys and − CH_3 could give different reaction with IAM. The derivatization of SeCys was fast and complete based on the occurrence of one narrow peak in the chromatogram by RP-ICP-MS and confirmation on ESI-MS of the CAM SeCys standard at mass to charge ratio (m/z) 227 showing the isotopic pattern of Se. ESI spectra of

CAM SeMet standard did not give the expected peak at 255 m/z, but rather a peak at 194, which may be explained by the loss of — CO_2 (44 m/z) and NH_3 (17 m/z) from CAM SeMet [43]. The RP-chromatograms of the muscle samples showed one distinct peak of SeMet identified by standard addition at 4 min, a smaller peak identified as protein bound SeCys after 2.7 min and four smaller unidentified Se species. In the liver samples CAM SeMet eluted in two peaks (retention time of ca. 5.4 and 5.9 min) and protein bound CAM SeCys at 4 min, as for the muscle samples four smaller unknown peaks were observed in the chromatograms.

3.2.4 Se amino acid concentration in liver and muscle

SeMet was the predominant Se amino acid in livers of the chickens receiving dietary treatments with SW or SY, whereas SeCys was the dominating Se amino acid in livers of chickens receiving SS as dietary supplement. These results are not in line with the findings of [57] were SeCys was predominant in liver tissue irrespective of treatment (with feed Se concentrations of 0.3 and 0.5 mg Se kg^{-1}). On the other hand in experiments with lamb [59] and beef cattle [59] receiving SY doses tend to result in SeMet compromising the greater proportion of total Se in the liver tissue which is in accordance with the result in the present work using elevated Se concentrations in the diet (**Figure 3**).

The different Se amino acid portioning may reflect a saturation of SeCys in the liver at higher SW or SY doses and subsequent change from incorporation into selenoproteins to the non-specific incorporation of SeMet into general liver tissue proteins [57].

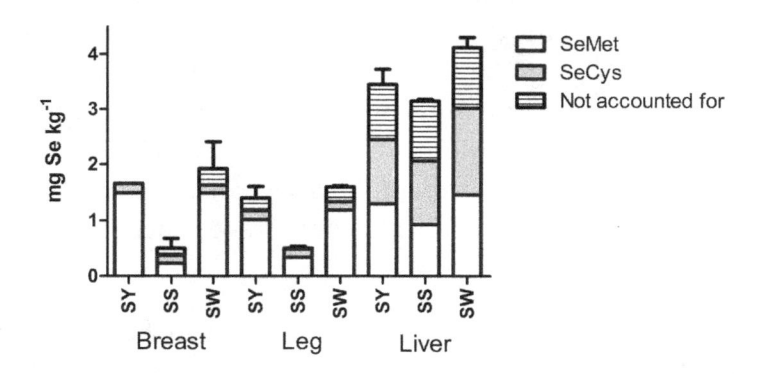

Figure 3. Concentration of Se amino acids in muscle and liver with regard to Se source (SY, Se-enriched yeast; SW, Se-enriched wheat and SS, sodium senenite).

The predominance of SeMet in breast tissue of chickens receiving SW or SY in the dietary treatment is in line with the observations of [42, 57].

The concentration of SeCys in breast and leg muscle tissues were irrespective of dietary treatment (0.15 ± 0.02 µg Se g^{-1}; $n = 6$) and in accordance with concentrations reported by [42], where 3 (0.11 ± 0.0, 0.18 ± 0.01 and 0.13 ± 0.08 µg Se g^{-1}) out of 4 (1.11 ± 0.35 µg Se g^{-1}) samples from chicken breast tissue had similar Se concentrations. It seems that the source of Se did not affect the concentration of SeCys in the muscle when there is an excess or adequate concentration of Se in the diet. The low level of SeMet found in tissues of chickens fed with selenite is likely due to other constituents in the diet, e.g. fishmeal and soybean [68]. The Se speciation analysis showed that the SeMet concentration was a factor 5 higher in chickens fed organic Se than in chickens fed with selenite. The present results show that Se from SS was less retained in muscle tissues compared to Se from SY and SW, and the retention observed is attributed to the incorporation as SeCys in proteins as stated by [65]. An increased organic Se source consisting primarily of SeMet would increase the SeMet concentration in the tissues and thereby the retention of Se in muscle tissue.

Even though the concentrations of Se in the liver were the highest, the effect on daily human intake would be of minor significance as the consumption of liver is low compared to the consumption of breast and leg muscle. Therefore focus should be on the increase of Se concentration in chicken breast and leg muscle by feeding organic Se to the chickens (SY and SW). The bioaccessibility of Se from the chicken meat to humans has previously been investigated by an in vitro model [41], showing relatively high accessibility of Se from chicken meat to humans (70–90% depending on the method used). Furthermore, human studies with isotopic labelled Se have demonstrated bioavailability over 90% from different food sources [69]. The Se biofortification of edible chicken meat products will increase the human Se intake and should be considered as a relevant strategy to increase the Se concentration in human plasma.

3.2.5. Amino acid composition

The use of organic and inorganic Se in the feed did not affect the overall amino acid distribution in the breast muscle tissues. The composition of amino acids in chicken meat is highly dependent on the composition of the diet. In this experiment the diets were added methionine (Met) (3 g kg^{-1}), lysine (Lys) (4 g kg^{-1}) and threonine (Thr) (1 g kg^{-1}), other important protein sources in the feed was soybean and fish meal. According to [63] a concentration of 3 g Met kg^{-1} will affect the bioavailability of SeMet and transfer to muscle, as the two amino acids are taken up by the same sodium transport system. This system is specific for neutral amino acids, resulting in inhibition of SeMet by Met as there is only trace concentration of SeMet compared to Met in the diet.

The amino acid distributions were compared to the Danish Food Composition Databank (DFCD) (Ed. 07.01) and What's In The Food You Eat Search Tool from United State Department of Agriculture, the amino acid composition of the breast muscles were comparable to what is reported in the DFCD but not from USA. This is probably because the Danish and Norwegian diets are wheat based whereas the diets in USA are maize based. The knowledge of the interaction between Se and Met is limited [63] and since all diets had the same Met concentrations it is impossible to extract any information on the influence on SeMet in this experiment.

3.2.6. Gene expression of muscle samples

Muscular gene expressions in *Gallus gallus* as a result of inorganic versus organic Se supplement (SS and SW) were determined with the use of a 44k Chicken oligonucleotide microarray. The results indicated similar expression of genes coding for selenoproteins, which is in line with the finding of equal SeCys concentrations in the muscles. However, global gene set functional enrichment analysis (Fatiscan) revealed statistically significant enrichment ($p < 0.05$) of a number of biological processes that were dependent on the Se feed sources, such as cell growth, organ development, protein metabolism (**Table 1**). Although Bayers statistics were not able to detect changes in differential expression of single genes, fatiscan analysis based on ranking of statistical values from these gene-centric analysis were able to identify subtle changes affecting a broad set of biological responses. Future effort to elucidate the biological meaning of the findings may potentially provide a mechanistic understanding to characterize molecular signatures associated with organic and inorganic Se supplements.

Functional term	Score*
Anatomical structure development	5.67
Organ development	4.34
Anatomical structure morphogenesis	2.79
Protein catabolic process	2.31
Nervous system development	2.18
Nitrogen compound metabolic process	2.18
Skeletal system development	1.96
Cellular nitrogen compound metabolic process	1.96
Regulation of cell size	1.96
Cell growth	1.6
Brain development	1.6

*A term annotation weight can be computed as the number of sequences annotated to that term or as an annotation confluence score. This confluence score (Node Score) takes into account the number of sequences converging at one gene ontology term and penalizes by the distance to the term where each sequence actually was annotated.

Table 1. Differences in regulations of genes in biological processes in chicken muscle of chickens fed SW ($n = 4$) compared to muscle of chickens fed SW ($n = 4$).

By comparing the fold change of genes involved in Se processes, it could be deduced that the use of Se-enriched wheat positively regulated developmental and metabolic processes in the muscle compared to selenite. The difference may be due to differences in uptake and bioavailability interacting (stimulatory/suppression) with biological processes. Brennan et al. [70] reported similar findings in oviducts of female chickens indicating differences in growth and metabolic patterns being higher when supplemented with SY compared to SS at lower Se concentrations (0.3 mg Se kg^{-1}).

4. Conclusion

Organic Se is three times more efficient than inorganic Se in increasing the chicken breast and leg muscle Se concentration, due to the high accumulation of SeMet in muscle proteins. The Se-source neither affects the muscle SeCys concentration nor the amino acid composition in the muscles. The use of SW as dietary supplement in chicken feed is an alternative to the SY based supplements to increase Se and in particular SeMet concentrations in edible chicken parts. The bioactive Se is defined as the fraction of Se from feed or food that is converted into biological active selenoproteins. SeCys is incorporated into all selenoproteins and quantification of SeCys will therefore give a measure of the bioactive amounts of Se in different tissues. A relatively high proportion of the Se from the diets were converted into bioactive SeCys in the liver ($35 \pm 2.2\%$), no significant difference between inorganic and organic Se sources. The concentration of SeCys in breast and leg muscle tissues were irrespective of dietary treatment. The results indicated similar expression of genes coding for selenoproteins, which is in line with the finding of equal SeCys concentrations in the muscles.

Due to longer retention time of SeMet in muscle, and possibly other organs, SeMet acts as a reservoir of Se in the body. The SeMet reservoirs could sustain the SeCys status of the chickens over time [57]. To our knowledge there is no known biological functions of SeMet, and the inorganic forms are readily transformed to SeCys through selenide in liver. One could argue that the chickens do not need the extra pool of Se in the form of SeMet in e.g. muscles, as the inorganic form will be sufficient to meet the chickens Se requirements. Since the feed is fortified with selenite the chickens will always have adequate access to the essential trace element. However, global gene set functional enrichment analysis revealed statistically significant enrichment of a number of biological processes that were dependent on the Se feed sources, such as cell growth, organ development and protein metabolism in favour of organic Se.

Selenium from wheat or chicken muscle could be important Se sources to ensure adequate Se intake in humans in Se sub deficient or deficient populations.

Acknowledgements

The authors would like to acknowledge Dr. Espen Govasmark and Dr. Katarzyna Bierla for their help with speciation analysis and Prof. Knut Erik Tollefsen with statistics and help with gene expression analysis.

Author details

Anicke Brandt-Kjelsen[1*], Brit Salbu[1], Anna Haug[2] and Joanna Szpunar[3]

*Address all correspondence to: anicke.brandt-kjelsen@nmbu.no

1 Department of Environmental Sciences, Norwegian University of Life Sciences (NMBU), Ås, Norway

2 Department of Animal and Aquacultural Sciences, Norwegian University of Life Sciences (NMBU), Ås, Norway

3 Laboratorie de Chimie Analytique Bio-inorganique et Environement, CNRS UMR 5254, Pau, France

References

[1] Johansson, L., Gafvelin, G., and Arner, E. S. J. (2005). Selenocysteine in proteins properties and biotechnological use. Biochimica et Biophysica Acta-General Subjects 1726, 1–13.

[2] Terry, N., Zayed, A. M., de Souza, M. P., and Tarun, A. S. (2000). Selenium in higher plants. Annual Review of Plant Physiology and Plant Molecular Biology 51, 401–432.

[3] Chen, G. S., and Metzenbe, R. l., (1974). Isolation and properties of selenomethionineresistant mutants of neurospora-crassa. Genetics 77, 627–638.

[4] Huber, R. E., and Criddle, R. S. (1967). Comparison of chemical properties of selenocysteine and selenocystine with their sulfur analogs. Archives of Biochemistry and Biophysics 122, 164–173.

[5] Lyons, M. P., Papazyan, T. T., and Surai, P. F. (2007). Selenium in food chain and animal nutrition: lessons from nature. Asian-Australasian Journal of Animal Sciences 20, 1135–1155.

[6] Van Metre, D. C., and Callan, R. J. (2001). Selenium and vitamin E. Veterinary Clinics of North America-Food Animal Practice 17, 373–402.

[7] Haug, A., Eich-Greatorex, S., Bernhoft, A., Hetland, H., and Sogn, T. (2008). Selenium bioavailability in chicken fed selenium-fertilized wheat. Acta Agriculturae Scandinavica Section A—Animal Science 58, 65–7.

[8] Hartikainen, H. (2005). Biogeochemistry of selenium and its impact on food chain quality and human health. Journal of Trace Elements in Medicine and Biology 18, 309–318.

[9] Låg, J., and Steinnes, E. (1978). Regional distribution of selenium and arsenic in humus layers of Norwegian forest soils. Geoderma 20, 3–14.

[10] Eich-Greatorex, S., Sogn, T. A., Ogaard, A. F., and Aasen, I. (2007). Plant availability of inorganic and organic selenium fertiliser as influenced by soil organic matter content and pH. Nutrient Cycling in Agroecosystems 79, 221–231.

[11] Sager, M. (2006). Selenium in agriculture, food, and nutrition. Pure and Applied Chemistry 78, 111–133.

[12] Pyrzynska, K. (2002). Determination of selenium species in environmental samples. Microchimica Acta 140, 55–62.

[13] Elrashidi, M. A., Adriano, D. C., Workman, S. M., and Lindsay, W. L. (1987). Chemical-equilibria of selenium in soils—a theoretical development. Soil Science 144, 141–152.

[14] Govasmark, E., and Salbu, B. (2011). Translocation and re-translocation of selenium taken up from nutrient solution during vegetative growth in spring wheat. Journal of the Science of Food and Agriculture 91, 1367–1372.

[15] Manafi, M. (2015) Comparison study of a natural non-antibiotic growth promoter and a commercial probiotic on growth performance, immune response and biochemical parameters of broiler chicks. Journal of Poultry Science, 52, 274–281.

[16] Cappa, J. J., Yetter, C., Fakra, S., Cappa, P. J., DeTar, R., Landes, C., Pilon-Smits, E. A. H., and Simmons, M. P. (2015). Evolution of selenium hyperaccummulation in Stanleya (Brassicaceae) as inferred from phylogeny, physiology and X-ray microprobe analysis. New Phytologist, 2, 583–595.

[17] Gissel-Nielsen, G., Gupta, U. C., Lamand, M., and Westermarck, T. (1984). Selenium in soils and plants and its importance in livestock and human nutrition. Advances in Agronomy 37, 397–460.

[18] Walter, E. D., and Jensen, L. S. (1963). Effectiveness of selenium and noneffectiveness of sulfur amino acids in preventing muscular dystrophy in turkey poult. Journal of Nutrition 80, 327–331.

[19] Scott M. L., Olson G., Krook L., and Brown W. R. (1967) Selenium-responsive myopathies of myocardium and of smooth muscle in young poult. Journal of Nutrition 91, 573–583.

[20] Cantor and Tarino (1982) Comparative effects of inorganic and organic dietary sources of selenium on selenium levels and selenium-dependent glutathione-peroxidase activity in blood of young turkeys. Journal of Nutrition 112, 2187–2196

[21] Karlsen, J. T., Norheim, G., and Froslie, A. (1981). Selenium content of norwegian milk, eggs and meat. Acta Agriculturae Scandinavica 31, 165–170.

[22] Surai, P. F. (2002). Selenium in poultry nutrition 2. Reproduction, egg and meat quality and practical applications. Worlds Poultry Science Journal 58, 431–450.

[23] Surai, P. F., Karadas, F., Pappas, A. C., Sparks, N. H. C., (2006). Effect of organic selenium in quail diet on its accumulation in tissues and transfer to the progeny. British Poultry Science 47, 65–72.

[24] Schrauzer, G. N., Surai, P. F., (2009). Selenium in human and animal nutrition: Resolved and unresolved issues. A partly historical treatise in commemoration of the fiftieth anniversary of the discovery of the biological essentiality of selenium, dedicated to the memory of Klaus Schwarz (1914–1978) on the occasion of the thirtieth anniversary of his death. Critical Reviews in Biotechnology 29, 2–9.

[25] European Food and Safety Authority (EFSA), 2011: Scientific Opinion on the safety and efficacy of selenium in the form of organic compounds produced by the selenium-enriched yeast Saccharomyces cerevisiae NCYC R645 (SelenoSource AF 2000) for all species. EFSA Journal 9 (6), 2279.

[26] European Food and Safety Authority (EFSA), 2012: Scientific Opinion on safety and efficacy of selenium in the form of organic compounds produced by the selenium-enriched yeast *Saccharomyces cerevisiae* NCYC R646 (Selemax 1000/2000) as feed additive for all species. EFSA Journal 10 (7), 2778.

[27] European Food and Safety Authority (EFSA), 2008: Selenium-enriched yeast as source for selenium added for nutritional purposes in foods for particular nutritional uses and foods (including food supplements) for the general population—Scientific Opinion of the Panel on Food Additives, Flavourings, Processing Aids and Materials in Contact with Food. EFSA Journal 766, 1–42.

[28] Aspila, P. (1991). Metabolism of selenite, selenomethionine and feed-incorporated selenium in lactating goats and dairy-cows. Journal of Agricultural Science in Finland 63, 9–73.

[29] Cozzi, G., Prevedello, P., Stefani, A. L., Piron, A., Contiero, B., Lante, A., Gottardo, F., and Chevaux, E. (2011). Effect of dietary supplementation with different sources of selenium on growth response, selenium blood levels and meat quality of intensively finished Charolais young bulls. Animal 5, 1531–1538.

[30] EFSA FEEDAP Panel (EFSA Panel on Additives and Products or Substances used in Animal Feed). (2014). Scientific Opinion on the safety and efficacy of dl-selenomethionine as a feed additive for all animal species. EFSA Journal 2014;12(2):3567, 20 pp. doi: 10.2903/j.efsa.2014.3567

[31] Pedrero, Z., and Madrid, Y. (2009). Novel approaches for selenium speciation in foodstuffs and biological specimens: a review. Analytica Chimica Acta 634, 135–152.

[32] Larsen, E. H., Lobinski, R., Burger-Meÿer, K., Hansen, M., Ruzik, R., Mazurowska, L., Rasmussen, P. H., Sloth, J. J., Scholten, O., and Kik, C. (2006). Uptake and speciation of selenium in garlic cultivated in soil amended with symbiptic fungi (mucorrhiza) and selenate. Analytical Bioanlytical Chemistry 385, 1098–1108.

[33] Larsen, E. H., Sloth, S., Hansen, M., and Moesgaard, S. (2003). Selenium speciation and isotope composition in 77Se-enriched yeast using gradient elution HPLC and ICP-MS-dynamic reaction cell-MS. Journal of Analytical Atomic Spectrometry 18, 310–316.

[34] Polatajko, A., Sliwka-Kaszynska, M., Dernovics, M., Ruzik, R., Encinar, J. R., and Szpunar, J. (2004). A systematic approach to selenium speciation in, selenized yeast. Journal of Analytical Atomic Spectrometry 19, 114–120.

[35] Pyrzynska, K. (1998). Speciation of selenium compounds. Analytical Sciences 14, 479–483.

[36] Gammelgaard, B., Jackson, M. I., and Gabel-Jensen, C. (2011). Surveying selenium speciation from soil to cell-forms and transformations. Analytical and Bioanalytical Chemistry 399, 1743–1763.

[37] Gammelgaard, B., Gabel-Jensen, C., Sturup, S., and Hansen, H. R. (2008). Complementary use of molecular and element-specific mass spectrometry for identification of

selenium compounds related to human selenium metabolism. Analytical and Bioanalytical Chemistry 390, 1691–1706.

[38] Dumont, E., Vanhaecke, F., and Cornelis, R. (2006). Selenium speciation from food source to metabolites: a critical review. Analytical and Bioanalytical Chemistry 385, 1304–1323.

[39] Encinar, J. R., Ouerdane, L., Buchmann, W., Tortajada, J., Lobinski, R., and Szpunar, J. (2003). Identification of water-soluble selenium-containing proteins in selenized yeast by size-exclusion-reversed-phase HPLC/ICPMS followed by MALDI-TOF and electrospray Q-TOF mass spectrometry. Analytical Chemistry 75, 3765–3774.

[40] Moreda-Pineiro, J., Moreda-Pineiro, A., Romaris-Hortas, V., Moscoso-Perez, C., Lopez-Mahia, P., Muniategui-Lorenzo, S., Bermejo-Barrera, P., and Prada-Rodriguez, D. (2011). In-vivo and in-vitro testing to assess the bioaccessibility and the bioavailability of arsenic, selenium and mercury species in food samples. Trac-Trends in Analytical Chemistry 30, 324–345.

[41] Brandt-Kjelsen, A., Govasmark, E., Vegarud, G., Haug, A., Szpunar, J., and Salbu, B. (2012). In vitro digestion of selenium from selenium-enriched chicken. Pure and Applied Chemistry 84, 249–258.

[42] Bierla, K., Dernovics, M., Vacchina, V., Szpunar, J., Bertin, G., and Lobinski, R. (2008). Determination of selenocysteine and selenomethionine in edible animal tissues by 2D size-exclusion reversed-phase HPLC-ICP MS following carbamidomethylation and proteolytic extraction. Analytical and Bioanalytical Chemistry 390, 1789–1798.

[43] Lipiec, E., Siara, G., Bierla, K., Ouerdane, L., and Szpunar, J. (2010). Determination of selenomethionine, selenocysteine, and inorganic selenium in eggs by HPLC-inductively coupled plasma mass spectrometry. Analytical and Bioanalytical Chemistry 397, 731–741.

[44] Reyes, L. H., Marchante-Gayon, J. M., Alonso, J. I. G., and Sanz-Medel, A. (2006). Application of isotope dilution analysis for the evaluation of extraction conditions in the determination of total selenium and selenomethionine in yeast-based nutritional supplements. Journal of Agricultural and Food Chemistry 54, 1557–1563,

[45] Heras, I. L., Palomo, M., and Madrid, Y. (2011). Selenoproteins: the key factor in selenium essentiality. State of the art analytical techniques for selenoprotein studies. Analytical and Bioanalytical Chemistry 400, 1717–1727.

[46] Bugel, S. H., Sandstrom, B., and Larsen, E. H. (2001). Uptake and retention of selenium from shrimps in man. Journal of Trace Elements in Medicine and Biology 14, 198–204.

[47] Kapolna, E., Laursen, K. H., Husted, S., and Larsen, E. H. (2012). Bio-fortification and isotopic labelling of Se metabolites in onions and carrots following foliar application of Se and Se-77. Food Chemistry 133, 650–657.

[48] Suzuki, K. T. (2005). Metabolomics of selenium: Se metabolites based on speciation studies. Journal of Health Science 51, 107–114.

[49] Suzuki, K. T., Doi, C., and Suzuki, N. (2006). Metabolism of Se-76-methylselenocysteine compared with that of Se-77-selenomethionine and Se-82-selenite. Toxicology and Applied Pharmacology 217, 185–195.

[50] Tsuji, Y., Suzuki, N., Suzuki, K. T., and Ogra, Y. (2009). Selenium metabolism in rats with long-term ingestion of Se-methylselenocysteine using enriched stable isotopes. Journal of Toxicological Sciences 34, 191–200.

[51] Finley, J. W. (2006). Bioavailability of selenium from foods. Nutrition Reviews 64, 146–151.

[52] Thiry, C., Ruttens, A., De Temmerman, L., Schneider, Y. J., and Pussemier, L. (2012). Current knowledge in species-related bioavailability of selenium in food. Food Chemistry 130, 767–784.

[53] Fairweather-Tait, S. J., Bao, Y. P., Broadley, M. R., Collings, R., Ford, D., Hesketh, J. E., and Hurst, R. (2011). Selenium in human health and disease. Antioxidants & Redox Signaling 14, 1337–1383.

[54] Lu et al. (2009).

[55] Fox, T. E., Atherton, C., Dainty, J. R., Lewis, D. J., Langford, N. J., Baxter, M. J., Crews, H. M., and Fairweather-Tait, S. J. (2005). Uptake of selenium from wheat, garlic, and cod intrinsically labeled with Se-77 and Se-82 stable isotopes. International Journal for Vitamin and Nutrition Research 75, 179–186.

[56] Payne, R. L., and Southern, L. L. (2005). Changes in glutathione peroxidase and tissue selenium concentrations of broilers after consuming a diet adequate in selenium. Poultry Science 84, 1268–1276.

[57] Juniper, D. T., Phipps, R. H., and Bertin, G. (2011). Effect of dietary supplementation with selenium-enriched yeast or sodium selenite on selenium tissue distribution and meat quality in commercial-line turkeys. Animal 5, 1751–1760.

[58] Yoon, I., Werner, T. M., and Butler, J. M. (2007). Effect of source and concentration of selenium on growth performance and selenium retention in broiler chickens. Poultry Science 86, 727–730.

[59] Juniper, D. T., Phipps, R. H., Ramos-Morales, E., and Bertin, G. (2009). Effects of dietary supplementation with selenium enriched yeast or sodium selenite on selenium tissue distribution and meat quality in lambs. Animal Feed Science and Technology 149, 228–239.

[60] Haug, A., Graham, R. D., Christophersen, O. A., and Lyons, G. H. (2007). How to use the worlds scarce selenium resources efficiently to increase the selenium concentration in food. Microbial Ecology in Health and Disease 19(4), 209–228.

[61] Bierla, K. (2008). Analytical approaches to determination of selenium, selenoamino acids and selenium metabolites in biological materials using elemental and molecular mass spectrometry. PhD-thesis.

[62] Krstic, B., et al. (2012). Options for the production of selenized chicken meat. Biological Trace Element Research 146(1) 68–72.

[63] Skrivan, M., et al. (2008). Effect of dietary selenium on lipid oxidation, selenium and vitamin E content in the meat of broiler chickens. Czech Journal of Animal Science 53(7) 306–311.

[64] Govasmark, E., Brandt-Kjelsen, A., Szpunar, J., Bierla, K., Vegarud, G., and Salbu, B. (2010). Bioaccessibility of Se from Se-enriched wheat and chicken meat. Pure and Applied Chemistry 82, 461–471.

[65] Combs, G. F. (2001). Selenium in global food systems. British Journal of Nutrition 85, 517–547.

[66] Rayman, M. P., Infante, H. G., and Sargent, M. (2008). Food-chain selenium and human health: Spotlight on speciation. British Journal of Nutrition 100, 238–253

[67] Nordic Nutrition Recommendations (2012). Integrating nutrition and physical activity. © Nordic Council of Ministers, Copenhagen 2004.

[68] Yoshida, S., Haratake, M., Fuchigami, T., and Nakayama, M. (2011). Selenium in seafood materials. Journal of Health Science 57, 215–224.

[69] Sloth, J. J., Larsen, E. H., Bugel, S. H., and Moesgaard, S. (2003). Determination of total selenium and Se-77 in isotopically enriched human samples by ICP-dynamic reaction cell-MS. Journal of Analytical Atomic Spectrometry 18, 317–322.

[70] Brennan, K. M., Crowdus, C. A., Cantor, A. H., Pescatore, A. J., Barger, J. L., Horgan, K., Xiao, R., Power, R. F., and Dawson, K. A. (2011). Effects of organic and inorganic dietary selenium supplementation on gene expression profiles in oviduct tissue from broiler-breeder hens. Animal Reproduction Science 125, 180–188.

An Overall View of the Regulation of Hepatic Lipid Metabolism in Chicken Revealed by New-Generation Sequencing

Hong Li, Zhuanjian Li and Xiaojun Liu

Abstract

In chickens, more than 90% of the *de novo* synthesis of fatty acids occurs in the liver; therefore, the liver metabolism has a critical effect on chicken development and egg laying performance. Although the physiological processes of liver lipid metabolism have been studied extensively in chicken, the underlying mechanisms and the roles of noncoding RNAs in the process remain ambiguous. Recently, we investigated the regulatory mechanism of hepatic lipid in chicken by new generation sequencing technology. Our results uncovered many genes, which play crucial roles in mammal lipid metabolism process, might have different biological functions in chicken. Some other genes which might play essential roles in chicken hepatic lipid metabolism were found. In addition, the physiological processes of hepatic lipid metabolism in chicken are regulated by noncoding RNAs, such as miRNAs and lncRNAs.

Keywords: lipid metabolism, new generation sequencing, miRNA, lncRNA, chicken

1. Introduction

The molecular regulatory mechanisms of the hepatic lipids in domestic chicken had been largely established after being extensively studied (see reviews [1–5]). In recent years, however, with research advances in genomics, epigenomics and related fields such as systematic biology and bioinformatics, and also with the development of advanced techniques such as new generation sequencing and computation programming, our knowledge about gene regulation and interactions has been considerably widened. As a result, the following questions about synthesis, formation and transport of yolk precursors in liver of laying hens remains to be fully elucidated.

First, what enzymes actually catalyze lipids synthesis in laying hen liver? In comparison with mammals, liver is the major site of lipid biosynthesis in the chicken [6–8]. Though most of the chicken genes and their products involved in the hepatic lipid metabolism are highly similar to those in mammals including human, their specialized tasks were considerably different [9–12]. For instance, a recent study on lysophosphatidylglycerol acyltransferase 1 (LPGAT1) indicated that LPGAT1 has a role in lipid synthesis in mice [13]. Our studies revealed, however, LPGAT1 has no significant effect on lipid synthesis with estrogen induction in chicken. In addition, some of the genes related to lipid metabolism had been lost in chicken during evolutionary process [14]. Therefore, the range of genes and their products involved in the hepatic lipid metabolism in laying hen remain to be fully elucidated [8].

Second, how are the very low density lipoprotein (VLDL) particles assembled and secreted in the liver of chicken? In mammals, it was well documented that microsomal triglyceride transfer protein (MTTP) assists in lipoprotein assembly to form very low density lipoprotein [13, 15–19]. The formation of VLDL particles in avian species is tightly regulated by estrogen. However, recent study has proved that upregulation of *MTTP* in the liver is not required for the increased VLDL assembly during egg production in the chicken [20]. Our study on *MTTP* expression levels in livers between pre-laying and egg-laying hens also showed no difference though that the liver *ApoB* and *ApoVLDL II* expression levels and plasma VLDL level were elevated dramatically in laying hen.

Third, how does the estrogen induce lipid synthesis and transfer processes in liver of laying hen regulated by noncoding RNA? It is now well appreciated that a large portion of the eukaryotic genome gives rise to non-protein-coding RNAs (ncRNAs) of various sizes ranging from ~20 nucleotides to ~100 kb, which are predicted to play essential roles in a variety of biological processes (see reviews [21–24]). Among ncRNAs, microRNAs (miRNAs) and long ncRNAs (lncRNAs) attracted more researches' attention.

MiRNAs are short, being composed of only 18–25 nucleotides (nt) single-stranded RNAs, which was first described in 1993 [25]. Since then, the view of gene expression regulation has been dramatically altered. MiRNAs are reported to regulate gene expression at the posttranscriptional level through RNA interference (RNAi) pathways [26]. In general, miRNAs interact with mRNAs to perform their functions. It has been argued that one miRNAs can regulate the expression of hundreds of mRNAs, while the expression of one mRNA could be regulated simultaneously by hundreds of miRNAs [27]. In other words, miRNAs can play critical roles through constructing networks of sophisticated regulatory systems in organisms [28]. Currently, many varieties of miRNAs are widely reported in plants, animals and even microbes. Alterations of specific miRNA levels have significant correlation with changes of physiological or pathological functions of divergent origin.

LncRNAs are RNA polymerase II (RNAPII) transcripts that are longer than 200 nucleotides [29, 30], which may regulate protein-coding gene expression at both the transcriptional and posttranscriptional levels. Transcription regulated by lncRNAs could negatively or positively control protein-coding gene expression either in cis or in trans [31]. Posttranscriptional regulation by lncRNAs could also negatively or positively control protein coding gene expression through competing endogenous RNAs, modulating mRNA stability and translation by homologous base pairing, or acting as nuclear retention of mRNAs [32]. A growing number of lncRNAs have recently been described, and their functions are been uncovering.

Therefore, the objective of this chapter is going to give an overview of the molecules regulation of hepatic lipid metabolism in chicken, which was based on the studies performed by using the new generation sequencing technology.

2. Genes involved in hepatic lipid metabolism in chicken

Liver as the most important metabolic organ where up to 90% of fatty acids are *de novo* synthesized in chicken [33–35]. It was found that the onset of laying in the poultry is preceded by large increase in the plasma-free fatty acids, lipids and phosphoproteins [36]. We used the pre-laying hens (20-week old) and egg-laying hens (30-week old) of Lushi green-shelled-egg chickens as the experiment model, and the most obvious physiological difference between the two stages is laying egg or not. Three pre-laying hens and three egg-laying hens, which were raised in cages under the same environmental conditions with *ad libitum* to food and water, were slaughtered. Liver tissues were harvested immediately and the RNA from the liver samples was extracted. The new generation sequencing technology was used to establish the gene expression profile [37]. Bioinformatic analysis methods were used to explore the genes involved in hepatic lipid metabolism, and uncover the regulatory mechanism of hepatic lipid metabolism in chicken.

In our research results, compared to pre-laying hen, there were 960 significant differentially expressed (SDE) genes obtained in the liver of egg-laying hen [37]. Among those SDE genes, many ones were enriched in lipid metabolism pathways (**Figure 1**).

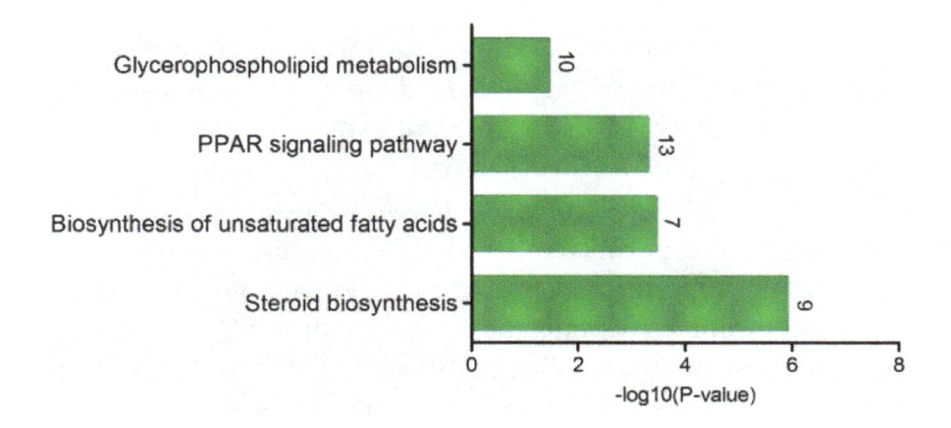

Figure 1. The SDE genes significantly enriched in lipid metabolism pathways. Note: The number on each bar means the number of genes enriched in the pathway.

For example, stearoyl-CoA desaturase 1 (SCD-1) is a rate limiting enzyme of monounsaturated fatty acid synthesis in liver and upregulated in egg-laying hens compared with pre-laying hens. Bioinformatic analysis showed it was enriched in the lipogenesis of the peroxisomeproliferator-activated receptor (*PPAR*) signaling pathway, and the mRNA expression and activity of *SCD-1* have been shown to be triggered by insulin to promote fat synthesis [38]. Interestingly, a very recent study demonstrated that B-cell translocation gene 1 (*BTG1*) overexpression inhibited the expression of *SCD-1* gene and altered hepatic lipid metabolism by decreased triglyceride accumulation in human [39]. However, our data showed that *BTG1* expression level also significantly increased when the *SCD-1* gene expression level elevated in egg-laying hens. It suggests that expression of *SCD-1* gene is regulated in different ways in chicken. The *FABP1* and *FABP3*, which are involved in hepatic fatty acid oxidation [40, 41], intracellular fatty acid transport [42], storage and export, as well as in cholesterol and phospholipid metabolism [43–45], were both significantly upregulated in the liver of egg-laying hens compared with pre-laying hens. They may promote lipid metabolism through the PPAR signaling pathway to meet the requirements of laying eggs. Some transcriptional factors such as sterol regulatory element binding protein (*SREBP-1*) and fatty acid synthase (*FASN*) genes were found to be elevated coordinately in egg-laying chicken liver that could synthesize fatty acids *de novo* [46]. Meanwhile, some novel genes and alternative splicing isoforms were also found to be differentially expressed and predicted to be relevant with lipid associated processes [37].

In the *de novo* fatty acids synthesis process, some key genes reported to be important in regulating lipid metabolism in mammals, but do not play the same roles in chicken (**Figure 2**). It is well documented that the triacylglycerol (TG) is postulated to synthesize through two biosynthetic pathways in liver. One is called glycerophosphate pathway, the other is monoacylglycerol pathway [47]. In the glycerophosphate pathway, TG is synthesized from the small precursor molecule glycerol-3-phosphate (G3P) and through the precursor phosphatidic acid (PA). The sequential reactions of acyl-CoA: G3P acyltransferase (GPAT) and acyl-CoA: 1-acyl-G3P acyltransferase (AGPAT) are involved in the incorporation of fatty acids into the glycerol backbone of phospholipids [47]. Glycerol-3-phosphate acyltransferase mitochondrial (GPAM) is an enzyme that plays a central role in *de novo* lipogenesis. The diacylglycerols (DG) is generated by PA dephosphorylation [48], and this process can be influenced by lipins [48, 49], which define a family of Mg^{2+}-dependent PA3 phosphatase enzymes with key roles in lipid metabolism [50]. Lipins have different expression patterns in different species, only one lipin in fungi, flies and worms [51], and three lipins including lipin1, 2 and 3 in mammals [52]. The DG can also be synthesized from monoacylglycerol (MG) catalyzed by Acyl-CoA:monoacylglycerol acyltransferase (MOGAT) family including MOGT1, MOGT2 and MOGT3 in mammals. In addition, LPGAT1 involves in triacylglycerol synthesis and secretion in liver [53] and promotes hepatic lipogenesis in mice [54]. In our study, compared to pre-laying hens, the expression levels *GPAM*, *AGPAT2*, *AGPAT3*, *lipin1* and *lipin2* genes were significantly upregulated in egg-laying hens. It suggested that these enzymes may play key roles in TG biosynthesis in the liver of chicken. However, some of the enzyme genes such as *GPAT2*, *AGPAT4*, *AGPAT5* and *AGPAT6* showed no changes in their expression levels, and some genes such as *AGPAT9*, *MOGAT1* and *LPGAT1* even exhibited down-regulated expression patterns. The other enzyme family members, which existed in mammals, were not detected in our animal model. Clearly, genes related to specific functions in regulating fatty acid synthesis are significantly different between mammals and avian species.

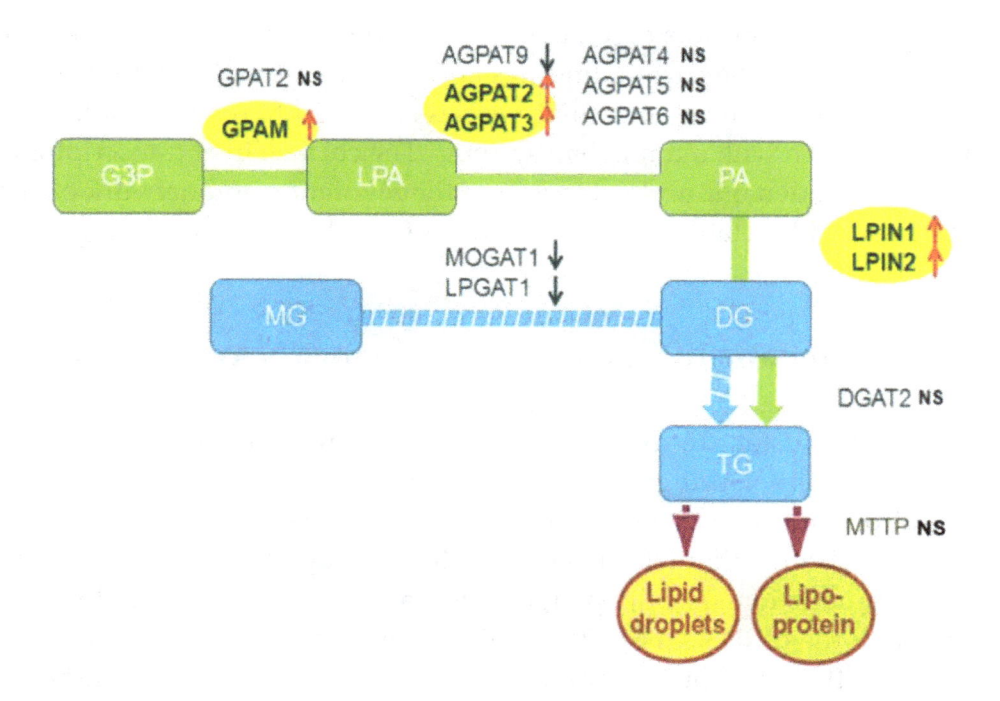

Figure 2. Expression pattern of key genes involved in chicken lipid metabolism. Note: NS means gene not significant differentially expressed in our RNA-seq results; up-arrow means gene up-regulated, down-arrow means down-regulated.

The final step in the *de novo* synthesis of TG is catalyzed by acyl-CoA: diacylglycerol acyltransferase (DGAT) enzymes, including DGAT1 and DGAT2 [55]. Overexpression of human *DGAT1* in McA-RH7777 cells can result in increasing the synthesis, accumulation and secretion of TG and VLDL [56]. Due to majority of the TG destined for secretion by liver is synthesized by *DGAT2* [57], expression of *DGAT2* in McA-RH7777 cells is positively related with the secretion of TG and apoB [56, 57]. However, the *DGAT1* gene was not expressed in chicken liver, and the expression level of *DGAT2* was not changed between the pre- and egg-laying hens. Another member of DGAT family, *SOAT1* (sterol O-acyltransferase 1) even was downregulated. It implied that there must be some other gene(s) involved in the process. Interestingly, a novel gene designated *DGAT2-like* gene, which possesses essential domains as does *DGAT2*, was identified and found to be significantly upregulated in egg-laying hens. This result suggests that *DGAT2-like* may play the role catalyzing TG formation in the liver of chicken as *DGAT2* does in mammals.

The MTTP assists in lipoprotein assembly to form low density lipoprotein [54, 58–62], and highly related with VLDL assembly and lipoprotein particle secretion [63, 64]. However, a previous study demonstrated that the upregulation of *MTTP* in liver was not required for increasing VLDL assembly during the laying period in chicken [20]. Same to the above result, our study also indicated that the *MTTP* was not significant differentially expressed in the liver of egg-laying hens in comparison to pre-laying hens. It implies that *MTTP–like* does not act the role as it does in mammals. As we expected, a novel gene-designated *MTTP-like*, which contains all the essential domains and motifs as *MTTP* does, was found to be significantly upregulated in egg-laying hens. Estrogen induction studies both *in vivo* and *in vitro* further revealed that the *MTTP-like* expression was regulated by estrogen in a dose dependent manner in liver of chicken. Although most the current findings appear to be consistent with the conservation of lipid metabolism in chicken and mammal, species-specific differences should be considered when comparing chicken with mammalian systems. The chicken liver transcriptome reported here could greatly broaden our understanding of the regulation and network of gene expression related to liver lipid metabolism in chicken at different physiological stages.

3. Regulation of hepatic lipid metabolism by ncRNAs in chicken

Lipid synthesis and transfer are dynamic and complex processes, which can be steered by various regulatory factors. During the egg-laying period, the estrogen level of hens goes up significantly and promotes the liver to synthesize egg yolk precursors. It was reported that estrogen can dramatically stimulate hepatic synthesis of apoB [65] and induce the *de novo* synthesis of the reproduction-specific apolipoprotein and apoVLDL-II in poultry by enhancing the accumulation of the mRNAs [66]. Our findings are consistent with previous reports that apoB and apoVLDL-II were significantly increased in the liver of egg-laying chicken compared with pre-laying hens (**Figure 3**). The increase in expression levels of *apoB, apoVLDL-II* and many other genes are supposed to be induced by estrogen. However, some upregulated genes such as sirtuin isoforms (*Sirt 1-7*) in egg-laying hens seems to be regulated by other factors instead of estrogen, because the expression levels of these genes in chicken liver tended to be decreased when chicken or chicken embryonic hepatic cells were treat with estrogen (unpublished data, related article is under reviewing).

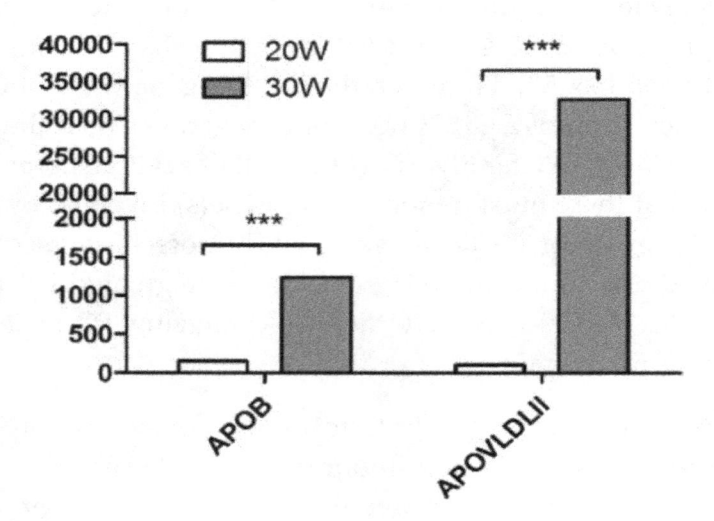

Figure 3. The expression of genes in 20- and 30-week-old hens.

MiRNA as a kind of posttranscriptional regulatory factor are reported to serve as important roles in lipid metabolism. It was identified that both gga-miR-148a and miR-122 are highly abundant miRNA in chicken hepatocytes [67] and in porcine liver [68]. A liver-specific miR-122 with high expression abundance in mammalian liver could modulate the hepatic fatty acids and cholesterol synthesis through repressing the expression of genes involved in cholesterol biosynthesis [69, 70]. MiR-33 involves in liver metabolism by regulating cholesterol efflux and high density lipoprotein metabolism by targeting the ATP-binding cassette subfamily A member 1 and ATP-binding cassette subfamily G member 1 [71]. These implied that some miRNAs may also involve in regulating chicken hepatic lipid metabolism through binding their target genes.

Considering the obvious difference of physiological activities between pre- and egg-laying stages, the pre- and egg-laying hens experiment model used in RNA-seq research [37] was used to investigate the critical miRNAs that may regulate the lipid metabolism. Bioinformatic analysis methods were used to explore the differentially expressed miRNAs involved in hepatic lipid metabolism and uncover the regulation ways of hepatic lipid metabolism in chicken [72]. Our results showed that majority of the target genes of down-regulated miRNAs significantly enriched in lipid metabolism-related processes, and enzyme activity, iron, vitamin binding molecular function (**Figure 4**). It is consistent with the event that eggs are rich in essential amino acids and fatty acids, as well as of some minerals and vitamins [73]. Our results suggest that the differentially expressed miRNAs may participate in chicken hepatic lipid metabolism through acting with their target genes.

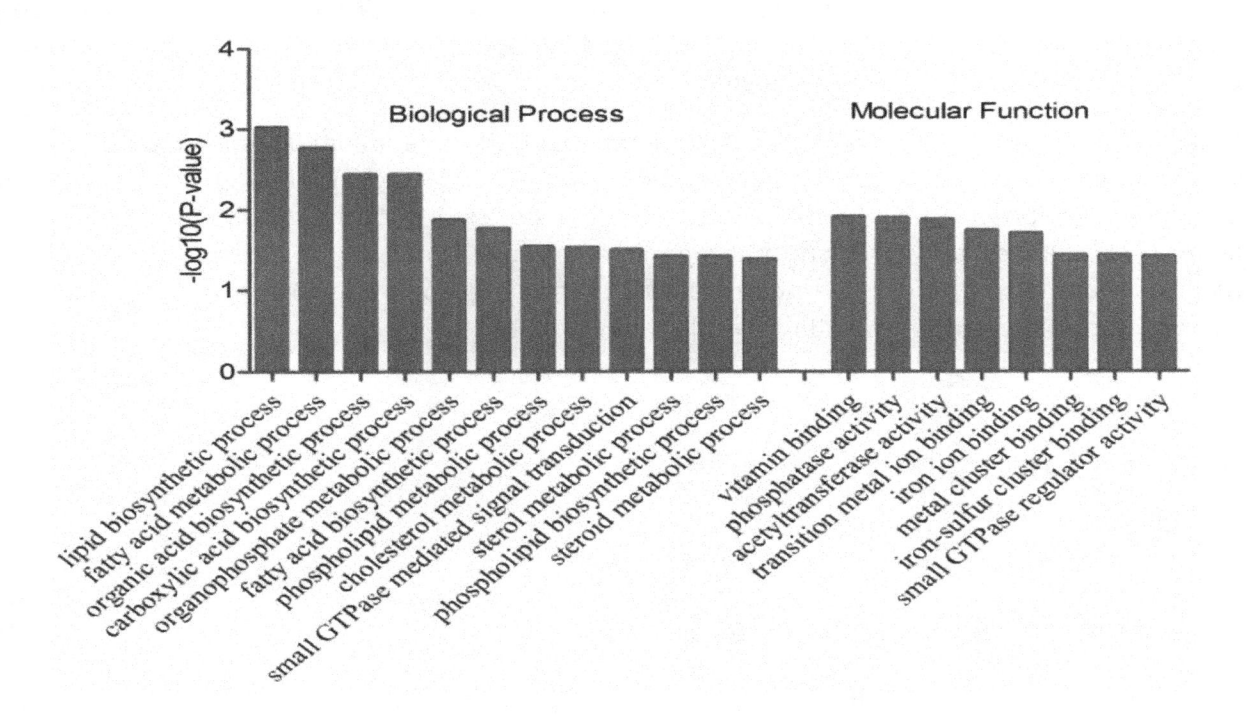

Figure 4. The significantly enriched and lipid-related GO terms of the target genes of the down-regulated miRNAs.

LncRNA is a class of pervasive genes involved in a variety of biological functions. Increasing researches present some lncRNAs are contributed to liver relevant metabolisms, including lipid metabolism. LncLGRAs, the transcriptional regulation factor of hepatic glucokinase (*GCK*) gene, can inhibit the expression of *GCK* and reduce hepatic glycogen content in mice during fasting [74]. It is reported that, whether enhanced the expression of lncRNA MALAT1 *in vivo* or *in vitro*, it can activate the nuclear *SREBP1c* expression and induce the intracellular lipid accumulation in mouse hepatocytes [75], while the lncRNAs that may take part in chicken hepatic lipid metabolism are unknown. Therefore, to gain insight into the underlying roles of lncRNAs serving as the hepatic lipid metabolism regulatory molecules, a lncRNA-Seq has be conducted to the livers of pre- and egg-laying hens to detect the lncRNAs.

4. Future perspectives

As well known, both lncRNAs and miRNAs serve as the endogenously expressed regulators of gene expression [76]. Recent researches have showed that the aberrant expression of lncRNAs and transcription factors can result in the miRNAs disorder. A study has demonstrated that a highly upregulated liver cancer lncRNA could serve as an endogenous sponge, which can down-regulate a series of miRNAs activities [77]. Due to the long size of lncRNAs, it regulate miRNA abundance via binding and sequestering them, working as the so-called miRNAs sponges, thus regulating the expression of target mRNAs [78, 79]. Given the complex modulation network among mRNAs [37], miRNAs [72] and lncRNAs, it will be great interest for us to combine these data sets to explore the possible regulation mechanisms among lncRNA, mRNA and miRNA. Our results will be a valuable resource for further elucidating the regulatory mechanism of chicken hepatic lipid metabolism and may also provide reference for understanding the molecular mechanisms in other poultry and mammalian species.

It has to be mentioned that the regulation of hepatic lipid metabolism in chicken described in this chapter is based on comparative studies between pre- and egg-laying hens, in which estrogen is supposed to be the main factor influencing lipid metabolism. In fact, many other factors such as feed additives and photogenic compounds may also play important roles in the lipid metabolism process, while the regulatory mechanism that genes involved in may not be the same as the present results [80–82].

Author details

Hong Li, Zhuanjian Li and Xiaojun Liu*

*Address all correspondence to: xjliu2008@hotmail.com

Henan Agricultural University, Zhengzhou, Henan, China

References

[1] Schneider WJ. Yolk precursor transport in the laying hen. Current Opinion in Lipidology. 1995;6(2):92–6.

[2] Bujo H, Hermann M, Lindstedt KA, Nimpf J, Schneider WJ. Low density lipoprotein receptor gene family members mediate yolk deposition. The Journal of Nutrition. 1997;127(5 Suppl.):801S–4S.

[3] Schneider WJ. Low density lipoprotein receptor relatives in chicken ovarian follicle and oocyte development. Cytogenetic and Genome Research. 2007;117(1–4):248–55.

[4] Schneider WJ. Receptor-mediated mechanisms in ovarian follicle and oocyte development. General and Comparative Endocrinology. 2009;163(1–2):18–23.

[5] Stein Y, Stein O. Serum lipoproteins and the liver, synthesis and catabolism. Hormone and Metabolic Research. 1974;Suppl.Ser.4:16–23.

[6] Waterman RA, Romsos DR, Tsai AC, Miller ER, Leveille GA. Effects of dietary carbohydrate source on growth, plasma metabolites and lipogenesis in rats, pigs, and chicks. Proceedings of the Society for Experimental Biology and Medicine Society for Experimental Biology and Medicine. 1975;150(1):220–5.

[7] Leveille GA, Romsos DR, Yeh Y, O'Hea EK. Lipid biosynthesis in the chick. A consideration of site of synthesis, influence of diet and possible regulatory mechanisms. Poultry Science. 1975;54(4):1075–93.

[8] Riegler B, Besenboeck C, Bauer R, Nimpf J, Schneider WJ. Enzymes involved in hepatic acylglycerol metabolism in the chicken. Biochemical and Biophysical Research Communications. 2011;406(2):257–61.

[9] Kirchgessner TG, Heinzmann C, Svenson KL, Gordon DA, Nicosia M, Lebherz HG, et al. Regulation of chicken apolipoprotein B: cloning, tissue distribution, and estrogen induction of mRNA. Gene. 1987;59(2–3):241–51.

[10] Wiskocil R, Bensky P, Dower W, Goldberger RF, Gordon JI, Deeley RG. Coordinate regulation of two estrogen-dependent genes in avian liver. Proceedings of the National Academy of Sciences of the United States of America. 1980;77(8):4474–8.

[11] Hermier D, Catheline D, Legrand P. Relationship between hepatic fatty acid desaturation and lipid secretion in the estrogenized chicken. Comparative Biochemistry and Physiology Part A, Physiology. 1996;115(3):259–64.

[12] Mason TM. The role of factors that regulate the synthesis and secretion of very-low-density lipoprotein by hepatocytes. Critical Reviews in Clinical Laboratory Sciences. 1998;35(6):461–87.

[13] Soh J, Iqbal J, Queiroz J, Fernandez-Hernando C, Hussain MM. MicroRNA-30c reduces hyperlipidemia and atherosclerosis in mice by decreasing lipid synthesis and lipoprotein secretion. Nature Medicine. 2013;19(7):892–900.

[14] Dakovic N, Terezol M, Pitel F, Maillard V, Elis S, Leroux S, et al. The loss of adipokine genes in the chicken genome and implications for insulin metabolism. Molecular Biology and Evolution. 2014;31(10):2637–46.

[15] Hussain MM, Rava P, Pan X, Dai K, Dougan SK, Iqbal J, et al. Microsomal triglyceride transfer protein in plasma and cellular lipid metabolism. Current Opinion in Lipidology. 2008;19(3):277–84.

[16] Hussain MM, Iqbal J, Anwar K, Rava P, Dai K. Microsomal triglyceride transfer protein: a multifunctional protein. Frontiers in Bioscience: A Journal and Virtual Library. 2003;8:s500–6.

[17] Hussain MM, Rava P, Walsh M, Rana M, Iqbal J. Multiple functions of microsomal triglyceride transfer protein. Journal of Nutrition and Metabolism. 2012;9:14.

[18] Berriot-Varoqueaux N, Aggerbeck LP, Samson-Bouma M, Wetterau JR. The role of the microsomal triglygeride transfer protein in abetalipoproteinemia. Annual Review of Nutrition. 2000;20:663–97.

[19] Rustaeus S, Lindberg K, Stillemark P, Claesson C, Asp L, Larsson T, et al. Assembly of very low density lipoprotein: a two-step process of apolipoprotein B core lipidation. The Journal of Nutrition. 1999;129(2S Suppl.):463S–6S.

[20] Ivessa NE, Rehberg E, Kienzle B, Seif F, Hermann R, Hermann M, et al. Molecular cloning, expression, and hormonal regulation of the chicken microsomal triglyceride transfer protein. Gene. 2013;523(1):1–9.

[21] Djebali S, Davis CA, Merkel A, Dobin A, Lassmann T, Mortazavi A, et al. Landscape of transcription in human cells. Nature. 2012;489(7414):101–8.

[22] Mercer TR, Dinger ME, Mattick JS. Long non-coding RNAs: insights into functions. Nature Reviews Genetics. 2009;10(3):155–9.

[23] Wilusz JE, Sunwoo H, Spector DL. Long noncoding RNAs: functional surprises from the RNA world. Genes & Development. 2009;23(13):1494–504.

[24] Pauli A, Rinn JL, Schier AF. Non-coding RNAs as regulators of embryogenesis. Nature Reviews Genetics. 2011;12(2):136–49.

[25] Lee RC, Feinbaum RL, Ambros V. The *C. elegans* heterochronic gene lin-4 encodes small RNAs with antisense complementarity to lin-14. Cell. 1993;75(5):843–54.

[26] Lagos-Quintana M, Rauhut R, Lendeckel W, Tuschl T. Identification of novel genes coding for small expressed RNAs. Science. 2001;294(5543):853–8.

[27] Krek A, Grün D, Poy MN, Wolf R, Rosenberg L, Epstein EJ, et al. Combinatorial microRNA target predictions. Nature Genetics. 2005;37(5):495–500.

[28] Wang X, Yu J, Zhang Y, Gong D, Gu Z. Identification and characterization of microRNA from chicken adipose tissue and skeletal muscle. Poultry Science. 2012;91(1):139–49.

[29] Derrien T, Johnson R, Bussotti G, Tanzer A, Djebali S, Tilgner H, et al. The GENCODE v7 catalog of human long noncoding RNAs: analysis of their gene structure, evolution, and expression. Genome Research. 2012;22(9):1775–89.

[30] Perkel JM. Visiting "noncodarnia". BioTechniques. 2013;54(6):301, 303, 304.

[31] Kornienko AE, Guenzl PM, Barlow DP, Pauler FM. Gene regulation by the act of long non-coding RNA transcription. BMC Biology. 2013;11:59.

[32] Yoon JH, Abdelmohsen K, Gorospe M. Posttranscriptional gene regulation by long noncoding RNA. Journal of Molecular Biology. 2013;425(19):3723–30.

[33] Leveille GA, O'Hea EK, Chakrabarty K. In vivo lipogenesis in the domestic chicken. Experimental Biology and Medicine. 1968;128(2):398–401.

[34] O'hea E, Leveille G. Lipogenesis in isolated adipose tissue of the domestic chick (Gallus domesticus). Comparative Biochemistry and Physiology. 1968;26(1):111–20.

[35] O'hea E, Leveille G. Lipid biosynthesis and transport in the domestic chick (Gallus domesticus). Comparative Biochemistry and Physiology. 1969;30(1):149–59.

[36] Heald P, Badman H. Lipid metabolism and the laying hen: I. Plasma-free fatty acids and the onset of laying in the domestic fowl. Biochimica et Biophysica Acta (BBA)-Specialized Section on Lipids and Related Subjects. 1963;70:381–8.

[37] Li H, Wang T, Xu C, Wang D, Ren J, Li Y, et al. Transcriptome profile of liver at different physiological stages reveals potential mode for lipid metabolism in laying hens. BMC Genomics. 2015;16(1):763.

[38] Lefevre P, Diot C, Legrand P, Douaire M. Hormonal regulation of stearoyl coenzyme-A desaturase 1 activity and gene expression in primary cultures of chicken hepatocytes. Archives of Biochemistry and Biophysics. 1999;368(2):329–37.

[39] Xiao F, Deng J, Guo Y, Niu Y, Yuan F, Yu J, et al. BTG1 ameliorates liver steatosis by decreasing stearoyl-CoA desaturase 1 (SCD1) abundance and altering hepatic lipid metabolism. Science Signaling. 2016;9(428):ra50–ra.

[40] Veerkamp J. Fatty acid-binding protein and its relation to fatty acid oxidation. Molecular and Cell Biology. 1993;123:101–6.

[41] Kaikaus RM, Sui Z, Lysenko N, Wu NY, de Montellano PO, Ockner R, et al. Regulation of pathways of extramitochondrial fatty acid oxidation and liver fatty acid-binding protein by long-chain monocarboxylic fatty acids in hepatocytes. Effect of inhibition of carnitine palmitoyltransferase I. Journal of Biological Chemistry. 1993;268(36):26866–71.

[42] Woodford J, Behnke W, Schroeder F. Liver fatty acid binding protein enhances sterol transfer by membrane interaction. Molecular and Cellular Biochemistry. 1995;152(1):51–62.

[43] Jefferson J, Powell D, Rymaszewski Z, Kukowska-Latallo J, Lowe J, Schroeder F. Altered membrane structure in transfected mouse L-cell fibroblasts expressing rat liver fatty acid-binding protein. Journal of Biological Chemistry. 1990;265(19):11062–8.

[44] Martin GG, Danneberg H, Kumar LS, Atshaves BP, Erol E, Bader M, et al. Decreased liver fatty acid binding capacity and altered liver lipid distribution in mice lacking the liver fatty acid-binding protein gene. Journal of Biological Chemistry. 2003;278(24): 21429–38.

[45] Atshaves BP, McIntosh AM, Lyuksyutova OI, Zipfel W, Webb WW, Schroeder F. Liver fatty acid-binding protein gene ablation inhibits branched-chain fatty acid metabolism in cultured primary hepatocytes. Journal of Biological Chemistry. 2004;279(30):30954–65.

[46] Gondret F, Ferré P, Dugail I. ADD-1/SREBP-1 is a major determinant of tissue differential lipogenic capacity in mammalian and avian species. Journal of Lipid Research. 2001;42(1):106–13.

[47] Yamashita A, Hayashi Y, Nemoto-Sasaki Y, Ito M, Oka S, Tanikawa T, et al. Acyltransferases and transacylases that determine the fatty acid composition of glycerolipids and the metabolism of bioactive lipid mediators in mammalian cells and model organisms. Progress in Lipid Research. 2014;53:18–81.

[48] Nadra K, de Preux Charles A-S, Médard J-J, Hendriks WT, Han G-S, Grès S, et al. Phosphatidic acid mediates demyelination in Lpin1 mutant mice. Genes & Development. 2008;22(12):1647–61.

[49] Brindley DN, Pilquil C, Sariahmetoglu M, Reue K. Phosphatidate degradation: phosphatidate phosphatases (lipins) and lipid phosphate phosphatases. Biochimica et Biophysica Acta (BBA)-Molecular and Cell Biology of Lipids. 2009;1791(9):956–61.

[50] Han G-S, Wu W-I, Carman GM. The Saccharomyces cerevisiae Lipin homolog is a Mg2+-dependent phosphatidate phosphatase enzyme. Journal of Biological Chemistry. 2006;281(14):9210–8.

[51] Péterfy M, Phan J, Xu P, Reue K. Lipodystrophy in the fld mouse results from mutation of a new gene encoding a nuclear protein, lipin. Nature Genetics. 2001;27(1):121–4.

[52] Donkor J, Sariahmetoglu M, Dewald J, Brindley DN, Reue K. Three mammalian lipins act as phosphatidate phosphatases with distinct tissue expression patterns. Journal of Biological Chemistry. 2007;282(6):3450–7.

[53] Hiramine Y, Emoto H, Takasuga S, Hiramatsu R. Novel acyl-coenzyme A: monoacylglycerol acyltransferase plays an important role in hepatic triacylglycerol secretion. Journal of Lipid Research. 2010;51(6):1424–31.

[54] Soh J, Iqbal J, Queiroz J, Fernandez-Hernando C, Hussain MM. MicroRNA-30c reduces hyperlipidemia and atherosclerosis in mice by decreasing lipid synthesis and lipoprotein secretion. Nature Medicine. 2013;19(7):892–900.

[55] Walther TC, Farese Jr RV. Lipid droplets and cellular lipid metabolism. Annual Review of Biochemistry. 2012;81:687.

[56] Liang JJ, Oelkers P, Guo C, Chu P-C, Dixon JL, Ginsberg HN, et al. Overexpression of human diacylglycerol acyltransferase 1, acyl-coa: cholesterol acyltransferase 1, or acyl-CoA: cholesterol acyltransferase 2 stimulates secretion of apolipoprotein B-containing lipoproteins in McA-RH7777 cells. Journal of Biological Chemistry. 2004;279(43):44938–44.

[57] Liu Y, Millar JS, Cromley DA, Graham M, Crooke R, Billheimer JT, et al. Knockdown of acyl-CoA: diacylglycerol acyltransferase 2 with antisense oligonucleotide reduces VLDL TG and ApoB secretion in mice. Biochimica et Biophysica Acta (BBA)-Molecular and Cell Biology of Lipids. 2008;1781(3):97–104.

[58] Hussain MM, Rava P, Walsh M, Rana M, Iqbal J. Multiple functions of microsomal triglyceride transfer protein. Nutrition & Metabolism. 2012;9(1):1.

[59] Hussain MM, Rava P, Pan X, Dai K, Dougan SK, Iqbal J, et al. Microsomal triglyceride transfer protein in plasma and cellular lipid metabolism. Current Opinion in Lipidology. 2008;19(3):277–84.

[60] Hussain MM, Iqbal J, Anwar K, Rava P, Dai K. Microsomal triglyceride transfer protein: a multifunctional protein. Front Biosci. 2003;8:s500–6.

[61] Berriot-Varoqueaux N, Aggerbeck L, Samson-Bouma M-E, Wetterau J. The role of the microsomal triglyceride transfer protein in abetalipoproteinemia. Annual Review of Nutrition. 2000;20(1):663–97.

[62] Rustaeus S, Lindberg K, Stillemark P, Claesson C, Asp L, Larsson T, et al. Assembly of very low density lipoprotein: a two-step process of apolipoprotein B core lipidation. The Journal of Nutrition. 1999;129(2):463S–6S.

[63] Gordon DA, Jamil H. Progress towards understanding the role of microsomal triglyceride transfer protein in apolipoprotein-B lipoprotein assembly. Biochimica et Biophysica Acta (BBA)-Molecular and Cell Biology of Lipids. 2000;1486(1):72–83.

[64] Gordon DA, Wetterau JR, Gregg RE. Microsomal triglyceride transfer protein: a protein complex required for the assembly of lipoprotein particles. Trends in Cell Biology. 1995;5(8):317–21.

[65] Kudzma DJ, Swaney JB, Ellis EN. Effects of estrogen administration on the lipoproteins and apoproteins of the chicken. Biochimica et Biophysica Acta (BBA)-Lipids and Lipid Metabolism. 1979;572(2):257–68.

[66] Chan L, Jackson R, O'malley B, Means A. Synthesis of very low density lipoproteins in the cockerel. Effects of estrogen. Journal of Clinical Investigation. 1976;58(2):368.

[67] Wang X, Yang L, Wang H, Shao F, Yu J, Jiang H, et al. Growth hormone-regulated mRNAs and miRNAs in chicken hepatocytes. PLoS One. 2014;9(11): e112896.

[68] Xie S-S, Li X-Y, Liu T, Cao J-H, Zhong Q, Zhao S-H. Discovery of porcine microRNAs in multiple tissues by a Solexa deep sequencing approach. PLoS One. 2011;6(1):e16235.

[69] Esau C, Davis S, Murray SF, Yu XX, Pandey SK, Pear M, et al. miR-122 regulation of lipid metabolism revealed by in vivo antisense targeting. Cell Metabolism. 2006;3(2): 87–98.

[70] Krützfeldt J, Rajewsky N, Braich R, Rajeev KG, Tuschl T, Manoharan M, et al. Silencing of microRNAs in vivo with 'antagomirs'. Nature. 2005;438(7068):685–9.

[71] Rayner KJ, Suárez Y, Dávalos A, Parathath S, Fitzgerald ML, Tamehiro N, et al. MiR-33 contributes to the regulation of cholesterol homeostasis. Science. 2010;328(5985):1570–3.

[72] Hong Li, Zheng Ma, Lijuan Jia, Yanmin Li, Chunlin Xu, Taian Wang, Ruili Han, Ruirui Jiang, Zhuanjian Li, Guirong Sun, Xiangtao Kang, Xiaojun Liu. Systematic analysis of the regulatory functions of microRNAs in chicken hepatic lipid metabolism. Scientific Reports. 2016; 6:31766.

[73] McNamara D, Thesmar H. Eggs. Egg Nutrition Center: Washington, DC, USA. 2005.

[74] Ruan X, Li P, Cangelosi A, Yang L, Cao H. A long non-coding RNA, lncLGR, regulates hepatic glucokinase expression and glycogen storage during fasting. Cell Reports. 2016;14(8):1867–1875.

[75] Yan C, Chen J, Chen N. Long noncoding RNA MALAT1 promotes hepatic steatosis and insulin resistance by increasing nuclear SREBP-1c protein stability. Scientific Reports. 2016;6:22640.

[76] Kaur P, Liu F, Tan JR, Lim KY, Sepramaniam S, Karolina DS, et al. Non-coding RNAs as potential neuroprotectants against ischemic brain injury. Brain Sciences. 2013;3(1): 360–95.

[77] Wang J, Liu X, Wu H, Ni P, Gu Z, Qiao Y, et al. CREB up-regulates long non-coding RNA, HULC expression through interaction with microRNA-372 in liver cancer. Nucleic Acids Research. 2010;38(16):5366–83.

[78] Ye S, Yang L, Zhao X, Song W, Wang W, Zheng S. Bioinformatics method to predict two regulation mechanism: TF–miRNA–mRNA and lncRNA–miRNA–mRNA in pancreatic cancer. Cell Biochemistry and Biophysics. 2014;70(3):1849–58.

[79] Cesana M, Cacchiarelli D, Legnini I, Santini T, Sthandier O, Chinappi M, et al. A long noncoding RNA controls muscle differentiation by functioning as a competing endogenous RNA. Cell. 2011;147(2):358–69.

[80] Li XL, He WL, Wang ZB, Xu TS. Effects of Chinese herbal mixture on performance, egg quality and blood biochemical parameters of laying hens. Journal of Animal Physiology and Animal Nutrition (Berl). 2016; doi:10.1111/jpn.12473.

[81] Gholami-Ahangaran M, Rangsaz N, Azizi S. Evaluation of turmeric (Curcuma longa) effect on biochemical and pathological parameters of liver and kidney in chicken aflatoxicosis. Pharmaceutical Biology. 2016;54(5):780–7.

Assessment of Maize *(Zea mays)* as Feed Resource for Poultry

Herbert K. Dei

Abstract

Maize, also known as corn (*Zea mays* L), has been recognised worldwide as a major energy feed ingredient in the diets of poultry. Its major nutritional limitation has been the low protein content and poor protein quality, which necessitates the use of expensive high-protein supplements or synthetic amino acids such as lysine in diets containing large proportion of maize. Therefore, extensive research has been conducted by maize breeders on the world maize germplasms collection with the aim of improving its nutritive value, particularly protein quality for monogastric animals. This chapter assesses the genetic upgrading of the nutritional quality of maize protein that culminated in the development of a new class of maize known as "Quality Protein Maize (QPM)". Various studies on the nutritionally improved maize for poultry as well as future challenges confronting maize utilisation in poultry production are highlighted.

Keywords: maize (*Zea mays*), energy, protein quality, nutritive value, poultry

1. Introduction

Poultry (avian species) have been recognised as affordable source of high-quality protein worldwide in the forms of meat and eggs. The poultry sector has been shown to become the world's largest meat sector by 2020 (**Figure 1**). Besides, the sector continues to record high global output of eggs (e.g. 70 million metric tonnes in 2014) as additional high-quality protein food.

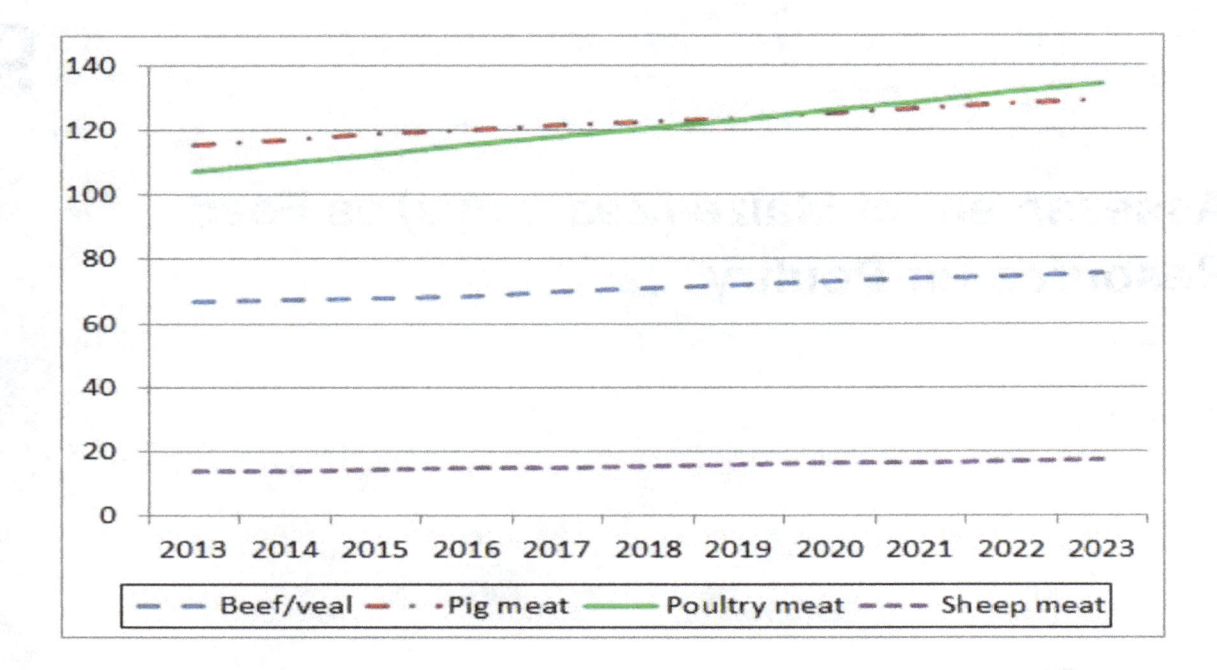

Figure 1. Global meat production (million metric tonnes) [1].

The rapid growth of the poultry sector is fuelled by several factors such as an increasing human population, greater purchasing power in developing economies, increased urbanisation and industrialisation in developing countries, development and transfer of feed, relatively short production cycle and advances in poultry breeding, and improved processing technologies. Of these factors, feed has been recognised as the most important factor controlling profitability and product quality [2].

Protein and carbohydrate are by far the two most important nutrients in poultry diets due not only to their marked effect on voluntary feed intake of the bird, but also the fact that they represent approximately 90% of the total cost of the ingredients in a ration [3]. Cereal grains constitute a large proportion (>50%) of poultry diets and contribute largely carbohydrates and to some extent proteins. They are mainly dietary source of energy, but can vary widely between grain types and animal species [4]. The common feed grains for poultry are corn or maize (*Zea mays*), wheat (*Triticum aestivum*), barley (*Hordeum vulgare*) and sorghum or milo (*Sorghum bicolor*).

Maize is by far the major feed grain grown worldwide, particularly in the United States. Although it is the preferred grain for feeding poultry [5], it is found to be low in protein content as well as protein quality [6], thereby limiting its nutritional value. This has necessitated a search for nutritionally improved maize varieties as well as alternative feed ingredients. The former has resulted in extensive research on the world maize germplasm collection with the aim of improving its nutritive value, particularly the protein quality for poultry.

This chapter discusses maize production and consumption, its genetic upgrading to improve the nutritional value with regard to feeding poultry.

2. Importance of maize in human and animal nutrition

Maize (*Zea mays* L.) tops other cereals in terms of worldwide production (**Figure 2**), with the United States being the largest producer as well as consumer (**Table 1**).

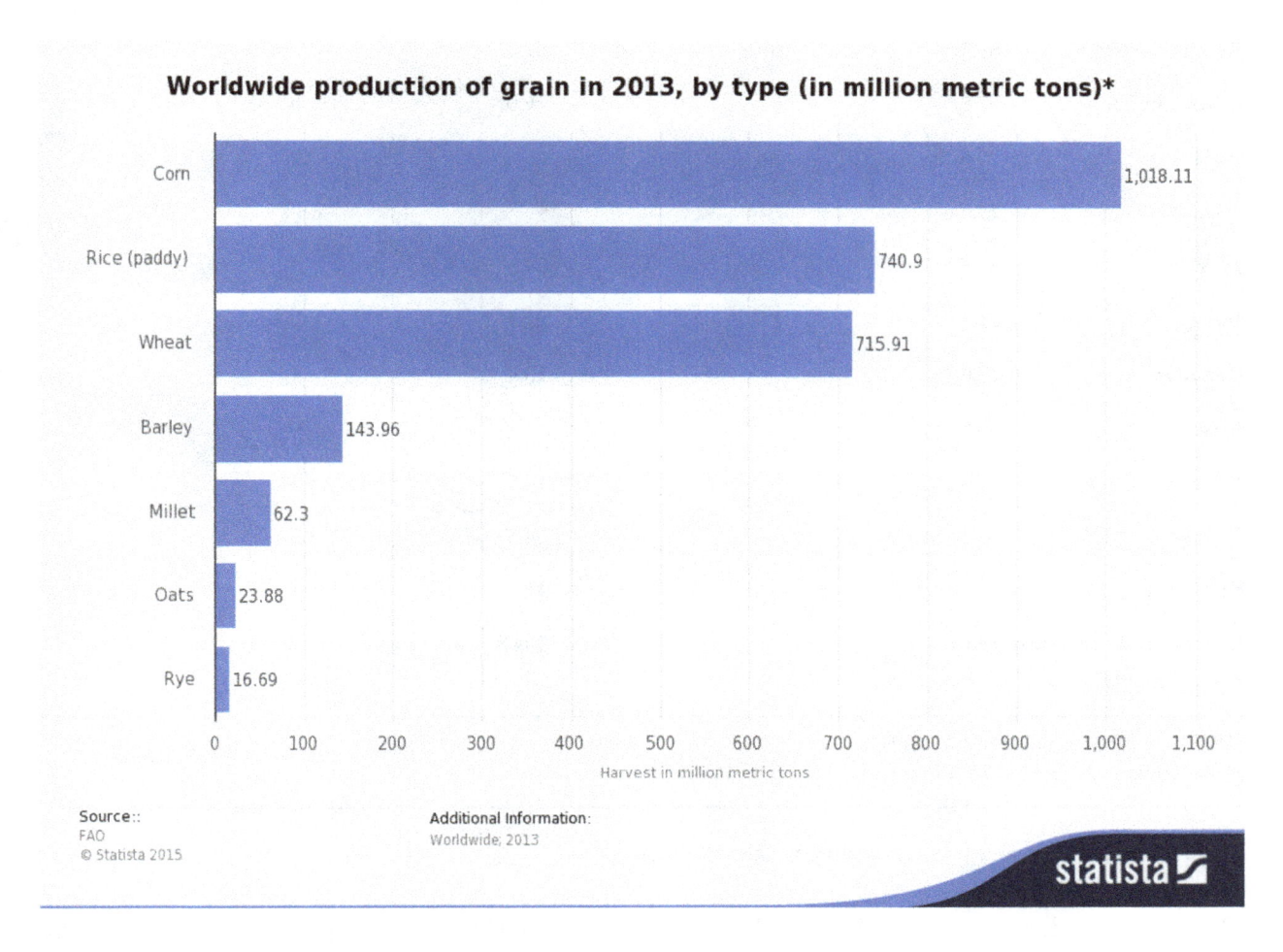

Figure 2. Global maize production [7].

On worldwide basis, much of the maize produced is fed to livestock, whereas only a small portion goes directly to human food [9]. The grain provides the world with 19% of its food calories and 15% of its annual production of food crop protein [6]. It is the basic staple cereal grain for large groups of people in Latin America, Africa and Asia [6], where the grain is consumed directly or in modified form as a major item of the diet.

Maize provides more feed for livestock than any other cereal grain [6]. For instance, 65% of the maize grown worldwide is used for livestock feed [10], of which the United States is the highest consumer. Also, rapid increase in poultry production in developing countries in Latin America, Africa and Asia is a major factor contributing to the increased use of maize for livestock feeding. In fact, maize is the preferred grain for feeding domestic birds, because its dietary energy value is the highest among cereals with very low variability between years for a given region [2].

Country	Production (million mt)[a]	~Production share (%)	Consumption (million mt)[b]	~Consumption share (%)
United States	345.5	36	301.5	31
China	224.6	23	214.0	22
Brazil	81.5	8	59.0	26
European Union-27	57.8	6	76.0	8
Argentina	25.6	3	–	–
Ukraine	23.5	2	–	–
Mexico	23.5	2	34.3	4
India	21.0	2	21.4	2
Canada	13.6	1	13.4	1
Russia	13.0	1	–	–
Japan	–	–	14.7	2
Egypt	–	–	14.5	2
Indonesia	–	–	12.7	1
Others including sub-Sahara Africa	138.4	14	204.8	21

[a]Total production (967.9 million mt).
[b]Total consumption (966.2 million mt).

Table 1. Worldwide maize production and consumption [8].

Over five hundred products [11] are obtained from industrial processing of maize, particularly from the main end-products of the "wet-milling" process of starch and nutritive sweeteners. The by-products include germ, bran and gluten which are suitable for feeding farm animals [12]. The gluten, in particular, is high in protein and metabolisable energy as well as a concentrated source of xanthophylls pigments, which make it popular in poultry production [2].

Besides, maize has long been an important ingredient in the manufacture of alcoholic beverages [9] including maize beer and whiskey. It is also an essential raw material in the production of industrial alcohols (25.4 kg of maize can yield 9.7 L of anhydrous ethanol plus useful by-products) [9]. The ethanol has a potential use as a partial replacement for gasoline due to increased fuel costs. The main by-product is referred to as "draff" or "distillers dried grains" (DDG), and it is high in protein. The DDG can be added another by-product called "solubles," which comprises the smallest residual particles of maize and yeast. The DDGS is high in protein, trace element and vitamins as well as increased availability of phosphorus, thereby making it popular feed ingredient for poultry production [2].

3. Nutritive value of normal hybrid maize grain

The maize grain (**Figure 3**) on dry matter basis is made up of 82.9% endosperm, 11.1% germ, 5.2% pericarp and 0.8% tip cap [13].

Table 2 shows the per cent chemical composition of the maize grain and grain fractions. In general, maize grain is low in protein content (9.1%), oil (4.4%) and ash (1.4%), but very high in starch content (73.4%) when considered on dry matter basis.

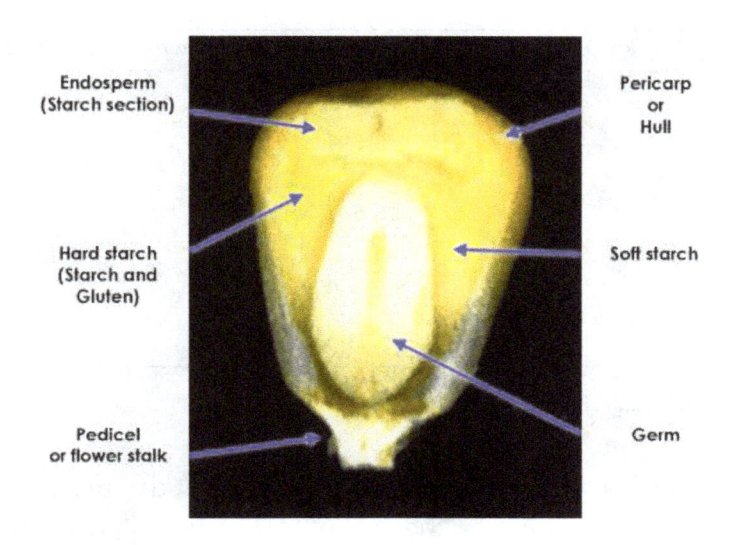

Figure 3. Structure of maize kernel (source: www.fao.org).

	Starch	Protein	Oil	Sugar	Ash (minerals)
Whole grain	73.4	9.1	4.4	1.9	1.4
Endosperm	87.6	8.0	0.8	0.6	0.3
Germ	8.3	18.4	33.2	10.8	10.5
Pericarp	7.3	3.7	1.0	0.3	0.8
Tip cap	5.3	9.1	3.8	1.6	1.6

Table 2. Chemical composition of normal maize grain and grain fractions (%DM) [14].

3.1. Carbohydrate content of maize grain

The relative proportions of the various carbohydrates are 77% starch, 2% sugars, 5% pentosans [15] and 1.2% crude fibre [16]. The carbohydrate which forms more than 70% of the maize grain [15] is concentrated in two starchy fractions, floury and flinty, of the endosperm. The sugar in the grain is found in the germ and dietary fibre is in the bran [13].

The endosperm consists of starch granules embedded in a protein matrix. Flinty endosperm has a more rigid protein structure and is also higher in protein content than floury endosperm

[13, 17]. The starch in the flinty endosperm consists of 100% amylopectin (large branched molecules), whereas that in the floury endosperm comprises about 27% amylose (linear molecules) and 73% amylopectin [16]. This variation in starch structure does not have any effect on the nutritional value of maize for poultry [2]. The distribution of flinty or floury endosperms in the maize grain determines whether a maize variety is classified as flint or floury (dent) maize. The starch which is the main source of energy in the grain has a digestible energy content ranging from 3.75 to 4.17 kcal/g dry matter [18], thereby making maize one of the highest in energy among the cereal grains (**Table 3**).

Cereal	Gross energy (MJ/kg DM)	Metabolisable energy (MJ/kg DM)
Barley	18.3	13.7
Sorghum	18.8	13.4
Maize	19.0	14.2
Millet	18.7	11.3
Wheat	18.4	14.0
Oats	19.0	11.5

Table 3. Gross and metabolisable energy of cereals [19].

The crude fibre content of maize grain averages about 2.7% of dry matter [20]. The crude fibre interferes with nutrient availability of the grain [21]. For instance, the range of protein digestibility of maize is 83–90% [21], while digestibility of carbohydrate is 99% [22]. Nevertheless, the maize grain is highly digestible due to its low crude fibre content [16].

3.2. Lipid content of maize grain

Maize oil (**Table 4**) is good quality oil both from nutritional standpoint (**Table 5**) and in terms of cooking quality [13]. Another desirable property of maize oil is its very low concentration of linolenic acid and high level of natural antioxidants [13, 23], thereby making the grain less susceptible to rancidity in storage.

Fatty acids	Structure[a]	Amount (%)
Palmitic	16:0	11.0
Stearic	18:0	3.0
Oleic	18:1	43.4
Linoleic	18:2	41.8
Linolenic	18:3	0.6

[a]Number of carbon atoms: number of unsaturated bonds.

Table 4. Concentration of fatty acids in maize oil triglycerides [23].

Grain fraction	Starch	Protein	Oil	Sugar	Ash
Endosperm	97.8	73.8	15.4	28.9	17.9
Germ	1.5	26.2	82.6	69.3	78.4
Pericarp	0.6	2.6	1.3	1.2	2.9
Tip cap	0.1	0.9	0.8	0.8	1.0

Table 5. Proportion of chemical constituents contained in each fraction of normal maize grain (% DM basis) [14].

The maize grain is a fair source of alpha-tocopherol (Vitamin E) which ranges from 0.6 to 2.1 mg/100 g grain [24]. Most of the carotenoids found in maize lipid are xanthophylls which are present only in yellow maize grain, and form about 12,511 mg/100 g grain [25]. Yellow maize, therefore, is one of the best sources of pro-retinal carotenoids [17]. These pigments cause yellow colouration of shanks and skin of broilers and yolks of eggs [26]. However, the pigmented grain tends to colour the carcass fat, which in the United Kingdom is not considered desirable [16].

3.3. Protein content of maize grain

The maize grain is deficient in protein, but its variability is low with standard error of the order 7 g/kg of crude protein [2]. The protein content of maize grain ranges from 8 to 11 g/100 g grain of dry matter [14, 23, 27]. The various fractions of grain vary considerably in protein content (**Table 5**). Even though the majority of protein in the grain occurs in the endosperm, the germ (184 g/kg DM) is considerably higher in protein content than the endosperm (80 g/kg DM) [14].

Generally, the low protein content of the grain limits its nutritive value as the only source of food for both humans and livestock.

The amino acid composition of whole maize grain is determined by both the relative proportions of the various protein fractions and the amino acid composition of each fraction [21]. Maize grain endosperm proteins are usually referred to as albumins, globulins, prolamins and glutelins, depending on their solubility in different solvent systems [21]. Prolamins and glutelins (also referred to as storage proteins) are confined to the endosperm, whereas albumins and globulins (also referred to as water-soluble proteins) are also found in the aleurone layer and the germ. The proportion of each protein fraction is presented in **Table 6**. In normal maize grain, the prolamin content exceeds that of glutelin and represents about 50–60% of the total protein [21]. Each protein fraction tends to have a characteristic amino acid composition (**Table 7**), and the relative proportion of each fraction strongly affects the level of individual amino acids in the total grain protein [28]. Prolamins are most deficient in lysine, thereby rendering maize protein poor in terms of nutritional quality. The general deficiency of lysine in maize grain is essentially the consequence of its low content of albumin and globulin, which besides having high lysine content exhibit a well-balanced amino acid composition similar to that of animal proteins of superior nutritional value [21]. Moreover, maize prolamins are characterised by larger quantities of leucine than isoleucine, thus causing the typical amino acid imbalance that further reduces the protein quality of maize [29].

Protein fraction	Amount (g/100 g protein)
Salt-soluble fraction (including NPN), i.e. albumins and globulins	6
Alcohol-soluble fraction (in the presence and absence of reducing agents), i.e. prolamins	64
Acid or alkali-soluble fraction (in the presence of reducing agents and nonextractable N), i.e. glutelins and residue	30

NPN, nonprotein nitrogen; N, Nitrogen.

Table 6. Protein fractions of normal maize [21].

	Protein fractions		
Amino acid	Albumin+globulin	Prolamin	Glutelin
Isoleucine	3.9–4.6	4.4–4.4	4.9–5.0
Leucine	5.4–6.4	20.3–20.5	9.4–10.1
Lysine	6.1–6.3	0.1–0.2	6.4–7.0
Cystine	0.3–1.6	0.1–0.7	0.2–0.5
Methionine	1.7–2.5	1.9–2.2	2.4–4.0
Tyrosine	2.8–4.5	5.5–6.1	2.4–2.7
Phenylalanine	2.4–4.2	7.8–8.0	5.3–5.6
Threonine	4.2–5.3	3.4–3.4	4.2–5.2
Valine	4.7–6.2	4.1–4.2	7.0–7.1

Table 7. Essential amino acid compositions of the endosperm protein fraction of normal maize (g/16 gN) [21].

The real significance of the poor nutritional quality of maize protein, therefore, is that the other food components of the diets of livestock and human may fail to provide adequate amounts of essential amino acids, particularly lysine to offset the nutritional deficiencies of maize protein [30].

3.3.1. Ways of improving the protein quality of maize

Although maize grain is relatively low in total protein and generally low in lysine and tryptophan, these shortcomings can be overcome by appropriate blending with animal products or legumes or oilseed products. The most obvious result of such blending is that the mixture is higher in protein than the maize component alone. Beyond this, animal products, legumes and various oilseed cakes improve the quality of maize protein by supplementing them with limiting amino acids such as lysine and tryptophan. This is called protein supplementation [31]. On the other hand, legumes and some oilseed cakes, which are deficient in methionine, can be supplemented by maize grain, which is not deficient in this amino acid. Such mutual balancing of each other's amino acids is known as protein complementation [31].

Besides, the quality of maize protein can be improved by the addition of synthetic amino acids like lysine. It also appears that a consistent enrichment of the nutritional quality of maize protein can be accomplished by developing new cultivars with a reduced content of prolamins with a parallel increase in glutelins and salt-soluble protein [21].

3.3.1.1. Protein supplementation of maize-based diets

Protein supplementation consists of adding small amounts of proteins which are rich sources of the amino acids deficient in normal maize [31]. Various protein supplements that have been tested on rats [32] include fish protein concentrate, soyabean flour, cottonseed flour, torula yeast, casein and egg protein. The recorded protein efficiency ratio values were normal maize (1.00), fish protein concentrate (2.44), soyabean flour (2.25) cottonseed flour (1.33), torula yeast (1.97), casein (2.21) and egg protein (2.24). This effect on protein quality was attributed to the contribution that the protein supplements made in lysine, tryptophan and protein content.

It has been reported that the supplementation of maize with soyabean flour increased usable protein from about 2.5% for maize alone to 10.6% when 20% of the soyabean flour was included in the diet for children [33, 34].

3.3.1.2. Protein complementation

Protein complementation comprises various food mixtures of maize with other ingredients of plant origin which are higher in protein quality and protein content [33, 35, 36]. One example of the favourable effect of protein complementation was established in rats [37].

Mixtures of cornflour and soyabean flour were fed to rats, and their weight gains per gram of protein consumed (protein efficiency ratio) were measured. Optimum results were obtained with the 40% cornflour and 60% soyabean flour ratio. With less soyabean flour in the mixture, lysine became limiting, but with more soyabean flour in the mixture, methionine was limiting (**Table 8**).

Maize:soyabean protein ratio	Amino acid content (g/16 gN)			Protein efficiency ratio
	Lysine	Total sulphur amino acids	Tryptophan	
100:0	2.88	3.15	0.60	1.6
80:20	–	–	–	2.3
60:40	–	–	–	2.7
40:60	4.95	3.14	1.07	2.9
20:80	–	–	–	2.8
0:100	6.32	3.12	1.38	2.6

Table 8. Complementation effects in rats fed combinations of soyabean flour and whole normal maize flour at a constant level of dietary protein [37].

The nutritive values of vegetable mixtures comprising 38% cottonseed flour plus 58% maize flour, 38% soyabean flour plus 58% maize flour, and 19% cottonseed flour plus 19% soyabean flour plus 58% maize flour with milk and egg proteins were compared in children [37]. The researchers reported biological values of the vegetable mixture which were close to those of high-quality reference proteins, milk and egg. The biological values registered were: milk (69%), egg (64%), soyabean plus maize (63%), cottonseed plus soyabean plus maize (53%) and cottonseed plus maize (50%). For normal maize alone, the biological value was 31%.

3.3.1.3. Amino acid supplementation of maize-based diets

The addition of synthetic amino acids such as L-Iysine and L-tryptophan either singly or in combination with maize-based diets to improve their protein quality has been demonstrated in feeding studies involving rats [38], pigs [39, 40], humans [34, 41] and poultry [42]. In all these trials, these amino acids have been found to improve protein quality of normal maize-based diets. **Table 9** shows substantial improvement in protein quality of normal maize-based diets supplemented with amino acids.

Amino acid added (g/kg)	Protein efficiency ratio of normal maize
None	1.21
Lysine (3), (1)	1.51
Tryptophan (0.5)	1.18
Lysine, tryptophan (3, 0.5); (1, 0.5)	2.66
Lysine, tryptophan, isoleucine (3, 0.5, 2.5)	2.58
Lysine, tryptophan, threonine (3, 0.5, 2)	2.56

Table 9. Biological confirmation of essential amino acid deficiencies in normal maize [43].

3.4. Vitamin content of maize grain

Vitamins in maize grain are concentrated mainly in the aleurone layer and the germ [13]. Analysis of the vitamin content of maize (**Table 10**) indicates that the grain furnishes significant quantities of riboflavin, panthothenic acid, choline and pyridoxine which are sufficient to satisfy the requirements of most livestock [44]. However, the most significant feature of the vitamin pattern in maize is the low niacin content. Besides, much of niacin that occurs in the grain is in a bound form (niacytin), which is not available to monogastric animals [13].

Furthermore, the high level of the essential amino acid, leucine, in the maize grain increases niacin requirement in humans [45]. Thus, people who live only on a diet of maize suffer from the disease pellagra, associated with niacin deficiency [6]. Nevertheless, niacin shortage alone would not cause pellagra if normal maize were rich in tryptophan [46] or heat-treated with alkali [30]. One approach for improving niacin intake in maize-based diet is complementation with either legumes or animal products [13].

Vitamin	Concentration (g/kg)
Carotene	4.6
Vitamin E	0.46
Thiamine (B_1)	4.83
Riboflavin (B_2)	1.61
Nicotinic acid	25.29
Pantothenic acid	6.44
Pyridoxine (B_6)	8.74
Choline	655.17

Table 10. Vitamin content of normal maize [16].

Yellow maize shows vitamin A activity, whereas white maize does not [47]. The vitamin A potency of yellow maize results primarily from the presence of carotenes in the grain. The carotene content of yellow maize is 0.46 mg/100 g of grain [16].

The occurrence of vitamins mainly in the aleurone layer and the germ implies that food preparations that do not retain these parts of the grain further decrease vitamins in the diet.

3.5. Mineral content of maize grain

The inorganic or mineral component (ash) of maize grain constitutes less than 2% [15]. Of this, about 75% is found in the germ. The grain is most abundant in phosphorus and potassium, but deficient in calcium and trace minerals except iron (**Table 11**). Much of the phosphorus, however, is present in the form of phytic phosphorus which is not digested by monogastric animals [48]. The little calcium that is normally present also has low bioavailability [13] because it forms complexes with phytic phosphorus.

Mineral	Concentration (mg/100 g)
Calcium	6.0
Phosphorus	300.0
Magnesium	160.0
Sodium	50.0
Potassium	400.0
Chlorine	70.0
Sulphur	140.0
Iron	2.5
Manganese	6.8
Copper	4.5

Table 11. Mineral content of normal maize grain [15].

3.6. Moisture content of maize grain

A moisture content of 10–14% is typical of properly ripened and dried maize grain [9]. The grain, therefore, furnishes a very high amount of dry matter. A moisture content of the grain higher than this may enhance the growth of moulds and cause the grain to rot in storage. Some of these moulds produce toxic metabolites like aflatoxins which can cause disease in humans and animals consuming the grain [9].

4. Factors affecting the chemical composition of maize grain

The factors that generally influence the chemical composition of maize grain are either genetic or environmental.

4.1. Genetic factors

Varieties of maize have been developed through breeding that contain up to 21% crude protein [49, 50–52]. However, maize varieties having more than 12% crude protein do have somewhat lower yields [51, 53], thus not suitable for commercial production.

New varieties of maize collectively called "high-lysine maize" contain nearly double the percentages of lysine and tryptophan of normal maize, even though the two types of maize are similar in overall protein content [6].

Maize varieties also differ in niacin content. For instance, inbred lines of dent maize have niacin content ranging from 13.9 to 53.3 µg/g of grain [44], whereas hybrids tend to be intermediate between their parent lines in concentration of this vitamin. Maize grain with a sugary endosperm had niacin content higher than that of waxy maize grain, which in turn had more niacin than dent grain [54].

The oil content of maize grain is largely characteristic of the particular variety. Breeding for high oil maize gave rise to strains of maize containing up to 20% oil in the grain [24, 55]. The existence of low oil maize varieties with average oil content of 1% [56] has been reported.

4.2. Environmental factors

Soil nitrogen appears to be the critical environmental factor that affects the protein content of maize [44]. Excess soil nitrogen beyond that required for maximum growth of the plant may increase the protein content of the grain [44]. It is reported that heavy application of nitrogen and phosphate fertilisers increased the thiamine content but decreased the level of niacin in maize grain [49].

Production year (rainfall and temperature) and location are often responsible for variation in the protein content of maize grains of the same variety of maize [52, 57, 58].

5. Maize grain as animal feed

As livestock feed, it is the grain that is most important. The stalks, leaves and immature ears are used as forage for ruminants [44]. Maize grain is recognised as giving the highest conversion of dry matter into meat, milk and eggs in relation to other cereal grains [13]. It is used extensively as the main source of calories in the feeding of poultry, pigs and cattle [6].

Maize grain has a digestible energy content of 3.75–4.17 kcal/g [18]. For chicken and pigs, the metabolisable energy values recorded when maize was fed were 3.6 and 3.8 kcal/g, respectively [18, 59], and corresponding gross energy digestibilities were 86% in chickens and 92% in pigs [18]. Maize, therefore, is popular for feeding monogastric animals, particularly poultry. For instance, maize is the basis of the high-energy poultry rations that are recognised throughout the United States whenever "broilers" are fattened [60].

In the feeding of poultry, maize grains are either fed directly or are milled and compounded with other ingredients and thoroughly mixed. The mixture is then fed or converted into forms most desired by specific animals.

The by-products obtained from both wet-milling and dry-milling industrial processes of maize grain are potential feed ingredients for poultry [12, 61–63] as depicted by the favourable nutrient composition of these by-products (**Table 12**) particularly in terms of protein content. The major by-product ingredients include the germ, bran and gluten [64]. These by-products of maize are usually mixed to produce a feed ingredient called maize gluten feed [62]. Despite the nutritional potential of maize gluten feed as a feedstuff for poultry, its use has been minimal due variously to paucity of research information available [65], perceived low metabolisable energy content [12, 66] and unknown quality of the protein [12] even though the protein content is fairly high.

	Moisture	Ash	CP	CF	EE	NFE	Ca	P
Maize feed meal	10.8	1.9	10.5	2.9	5.5	68.6	0.04	0.38
Maize bran	10.0	2.1	10.0	8.8	6.6	62.5	O. 03	0.14
Maize germ meal	7.0	3.8	20.8	7.3	9.6	51.5	0.05	0.59
Maize gluten meal	8.0	2.2	43.0	3.7	2.7	40.4	0.1	0.47
Maize gluten feed	9.5	6.0	27.6	7.5	3.0	46.4	0.11	0.78
Maize oil meal	8.7	2.2	22.1	10.8	6.8	49.4	0.06	0.62

Table 12. Chemical composition (%) of feedstuffs from maize and maize by-products [67].

6. Nutritionally improved maize grain

It has been known since 1914 that the quality of maize proteins is poor because they are deficient in the essential amino acids, lysine and tryptophan [38]. These deficiencies were attributed to the high zein fraction of maize protein in the maize grain of most varieties [68, 69]. Results obtained from extensive studies of zein indicated that it contains very low levels of lysine and tryptophan [70, 71]. Several researchers studied the factors that affected the protein quality of maize and reported that both the variety of maize and the environment had in several cases, a significant effect on lysine content [69, 72, 73]. It has been shown that the *opaque-2* gene in maize caused a genetic increase in lysine concentration of maize protein [74]. These researchers further reported that the lysine increment in *opaque-2* maize was the result of a change in the distribution of endosperm protein fractions, of which the *opaque-2* maize contained approximately 22% zein compared with 50% zein in normal maize. Chemical analysis of maize protein for amino acids [74–76] showed that *opaque-2* maize contained 60–130% more lysine than did normal maize, plus a 12–40% reduction in leucine as well as elevated level of tryptophan. Since these findings, several other mutant genes of maize have been identified. Collectively designated "high-lysine" genes, all of them control the level of zein accumulation during endosperm development. These "high-lysine" genes include most importantly *floury-2* [77]; *opaque-7* [78]; *opaque-6* and *floury-3* [79]. Of these genes, *opaque-2* has proven superior in zein reduction [80, 81].

The development of these nutritionally improved maize varieties is of particular significance to those who rely on maize as basic food and animal feed, and can thereby improve such diets nutritionally at no added cost.

6.1. Shortcomings of *opaque-2* gene as tool for improving protein quality in maize

Although the *opaque-2* gene favourably alters the amino acid spectrum in maize, it has several shortcomings that limit its widespread commercial use.

6.1.1. Grain yield

In general, *opaque-2* maize varieties have 10–15% lower grain yields than do normal maize varieties [82, 83]. Besides, *opaque-2* maize grain is 10–15% lighter in kernel weight [82, 83]. The lower grain weight can be attributed to loose packing of starch particles in the endosperm [84].

6.1.2. Moisture content of grain

Mature *opaque-2* maize grains are higher in moisture content by 1.8–4.2% than their comparable normal maize grains [80, 82]. Higher moisture content of the grains requires additional drying after harvest.

6.1.3. Grain appearance

Opaque-2 maize kernels are chalky and dull in contrast to the hard and shiny kernels of normal maize varieties [6]. The soft endosperm in *opaque-2* grains coupled with the dull appearance restricts acceptance by farmers, millers and consumers.

6.1.4. Susceptibility of grains to pests and diseases

Opaque-2 maize grains have been found to be more vulnerable to attack by *Sitophilus oryzea* [85] than normal type grains, in terms of both infestation and loss in weight of grains. *Opaque-2* varieties have also been reported to be more susceptible to *Chilo zonellus* [86]. Besides, *opaque-2* grains are more susceptible to seed rot caused by fungi such as *Cephalosporium acremonium* and *Fusarium moniliform* [87].

7. Development of quality protein maize (QPM)

Extensive field trials have been carried out at the International Center for Maize and Wheat Improvement (CIMMYT) in Mexico to identify the most productive *opaque-2* maize cultivars which are high in lysine and tryptophan contents as well as to change the soft endosperm in *opaque-2* grain into a conventional hard vitreous type [88–90]. Through backcrossing and several cycles of recurrent selection of maize, CIMMYT's maize breeders have successfully combined the high-lysine potential of the *opaque-2* gene with genetic endosperm modifiers. These new genotypes, collectively called "quality protein maize" (QPM), are becoming of major interest to seed producers, breeders, geneticists and industrialists for their large-scale production and for their potential advantages in human nutrition and animal feeding. The QPM grains are indistinguishable from normal types except by chemical analysis [6].

QPM cultivars retain the protein quality of conventional *opaque-2* maize but have improved agronomic traits, notably high yields and hard grain endosperm [6, 23, 81, 90–92].

Several experimental QPM cultivars yielded grains equal to those of the latest experimental releases of normal maize [6, 81, 90–92] in several regions of the world. Mexican QPM cultivars tested at more than twenty locations around the world have shown yields fully comparable to those of normal maize [93].

QPM grains are shiny, transparent and as hard as those of normal maize [6, 23, 94]. The QPM grain now has mostly the same density as that of normal maize, 1.29 g/cm^3 [6]. Grain sizes of both QPM and normal maize are similar [6]; however, some of the new QPM hybrids have a grain size greater than that of normal maize [23].

The moisture content of QPM grains at harvest is essentially identical to that of normal maize [6, 91]. On average, the QPM cultivars show higher incidence of ear rots but not disastrously so in humid regions [6]. Apart from ear-rotting organisms, other maize diseases seem to attack both QPM and normal maize with comparable severity [6].

As a result of the hard endosperm of QPM grains, the excessive incidence of broken grains and the accompanying storage damage have been eliminated. Thus, insect infestation is reduced and no worse than in normal maize [6].

7.1. Nutrient levels in QPM

The nutrient composition of QPM is similar to that of normal maize with the exception of lysine, tryptophan, leucine and isoleucine contents [6]. The protein content of QPM grains ranges from 7.4% to 10.5% of dry matter [23, 81, 90, 95–97], which is about the same as that of normal maize [6, 81].

The lysine and tryptophan contents of QPM grains are about twice those of normal maize. There is also reduced imbalance between isoleucine and leucine. The grains of most available QPM cultivars contain on the average 3.5–4.5% lysine of total protein in the grain [6, 81]. Also, the lysine content of the gluten of QPM grains is higher than that of normal maize grains [23].

The starch yielded by QPM grains is comparable to normal maize grains [98]. Starch contents of 56.6% grain for QPM and 55.0% grain for normal maize have been reported [23].

The fat content of QPM grains ranges from 3% to 7% [23, 94, 99]. Of the triglycerides, QPM grains contain significantly more palmitic acid and linoleic acid than normal maize grain but less of stearic, oleic and linolenic acids [23]. Yellow grain types of both QPM and normal maize contain similar levels of carotenoids [6].

8. Evaluation of protein quality of nutritionally improved maize

Many nutritional studies have demonstrated the benefits for monogastric animals of the high protein quality of nutritionally improved maize varieties collectively called "high-lysine maize" (of which *opaque-2* maize and QPM are outstanding) as compared with normal maize.

A summary of nitrogen balance studies with weanling rats to evaluate the protein quality of normal maize, *opaque-2* maize and QPM is presented in **Table 13**. The data indicate that *opaque-2* maize and QPM show significantly higher nutritional values than normal maize. The nutritional values are similar for *opaque-2* maize and QPM. The high efficiency of protein utilisation of high-lysine maize is explained on the basis of its good protein quality [95, 99–101].

Maize variety	Lysine (g/kg protein)	Biological indicator of nutritional quality			
		True digestibility (%)	Biological value (%)	Net protein utilisation (%)	Protein efficiency ratio
Normal maize[a]	27	–	62	55	1.5
Opaque-2 maize[a]	45	–	87	70	2.8
Normal maize[b]	–	–	–	–	1.6
Opaque-2 (soft)[b]	–	–	–	–	2.8
Opaque-2 (hard)[b]	–	–	–	–	2.9
Normal maize[c]	26	98	63	62	–
Opaque-2 maize	42	96	78	75	–
QPM[c]	40	97	76	74	–
Normal maize[d]	23	–	–	–	0.7
QPM[d]	36	–	–	–	1.3

Sources: [a]Bressani et al. [100];
[b]Mertz et al. [101];
[c]Villegas et al. [95];
[d]Ahenkorah et al. [99].

Table 13. Nutritional evaluation in rats of normal maize, *Opaque-2* maize and QPM.

Numerous feeding trials were carried out on rats to study the effect of protein quality of normal maize, *opaque-2* maize and QPM on weight gain, feed intake as well as efficiency of feed utilisation. Some of the published work are summarised in **Table 14**. These growth studies highlighted the superior feed value of QPM over normal maize. One of the nutritional benefits of improved protein quality of maize is a higher food intake from diets made up of the high-lysine maize as compared with normal maize diets [30]. This is because a diet of good protein quality stimulates food intake and a higher diet intake raises calorie intake [36, 100]. Also, efficiency of feed utilisation by rats fed high-lysine maize is better than for normal maize indicating the superior quality of high-lysine maize. The performance of rats fed high-lysine maize was only slightly lower than that observed for their counterparts fed casein (**Table 14**).

Diets	Average gains (g)	Average feed intake (g)	Feed/gain	Sources
Normal maize	25[a]	248[a]	9.92[a]	[100]
Opaque-2 maize	130[b]	455[b]	3.50[b]	
Casein	132[b]	408[b]		
Normal maize	50[a]	211[a]	4.22[a]	[102]
Opaque-2 maize	122[b]	281[b]	2.20[b]	
Normal maize	15[a]	131[a]	9.4[a]	[103]
QPM	27[b]	162[b]	7.0[b]	
Normal maize	9[a]	115[a]	12.78[a]	[99]
QPM	18[b]	154[b]	8.56[b]	
Casein	28[c]	156[b]	5.57[c]	

Values in the same column with different superscripts are statistically different (P<0.01).

Table 14. Comparative average weight gains, average feed intakes and feed conversion efficiencies in rats fed normal maize, high-lysine maize and casein.

8.1. Evaluation of nutritionally improved maize in poultry diets

In initial studies with young chicks [104, 105], only slight differences in growth rate and feed conversion efficiency of chicks fed *opaque-2* maize in place of normal maize in maize-soyabean meal diet which contained a dietary protein level of 15%. However, when a deficiency of the first limiting amino acid in this type of diet, methionine was corrected by supplementation, significantly better gains and feed conversions were observed with *opaque-2* maize as compared with normal maize. Since supplementation of normal maize diets with lysine up to the level in *opaque-2* maize diets resulted in equal performance, it was concluded that the beneficial effects of *opaque-2* maize over normal maize were mediated solely through the higher lysine content of *opaque-2* maize. Analyses of plasma amino acids revealed higher lysine levels in chicks fed *opaque-2* maize as compared with chicks fed normal maize. It was reported that chicks given high-lysine maize diets at a dietary protein level of 15% out-performed their counterparts fed normal maize diets at the same dietary protein level [106]. Again, it was found that *opaque-2* corn had higher nutritive value than normal corn as determined in modified chick bioassay of protein efficiency ratio [107].

Other reports that demonstrated the superior feeding value of high-lysine maize are summarised in **Tables 15–17**. Also, it was reported that chicken fed QPM exhibited improved growth rate by 20% as compared with those fed normal maize [92]. The data showed that there was no response to tryptophan supplementation of high-lysine maize. This means that the amount of tryptophan in high-lysine maize is sufficient to meet the requirements of chicks. On the other hand, a significant response is obtained when normal maize is supplemented with tryptophan. Thus, high-lysine maize is superior to normal maize not only for its higher lysine content but also for its higher tryptophan content. Furthermore, there is positive response to methionine supplementation of high-lysine maize as previously reported [104, 105], making the protein quality of the high-lysine maize superior to normal maize for the chick. In addition, the results from the substitution of high-lysine maize protein for soyabean or fishmeal protein indicate that high-lysine maize protein is comparable in quality to that of fishmeal or soyabean meal for the chick. Another interesting observation from these data is that lysine supplementation of high-lysine maize diets resulted in improved performance comparable to that of chicks fed control diets. This indicates that high-lysine maize alone is not adequate as a sole source of protein or lysine for chicks [6].

		Two-week-old chicks	
Source of protein	Protein level (%)	Average body weight (g)	Protein efficiency ratio
Basal protein+SBM	14	124.7[a]	2.99[a]
Basal protein+NM	14	77.7[d]	2.31[c]
Basal protein+OP (hard)	14	81.6[c]	2.45[b]
Basal protein+OP (soft)	14	92.2[b]	2.81[a]
Basal protein+Fl-2	14	81.2[c]	2.46[b]
BP+SBM+Lys (3 g/kg)	14	122.8[c]	3.24[a]
BP+NM+Lys (3.2 g/kg)	14	86.6[c]	2.74[c]
BP+OP (hard)+Lys (3 g/kg)	14	114.0[a]	3.20[a]
BP+OP (soft)+Lys (2 g/kg)	14	116.0[a]	3.15[ab]
BP+Fl-2+Lys (8 g/kg)	14	98.5[b]	2.94[b]

BP: basal protein; Fl-2: *floury-2* maize; OP: *opaque-2* maize; SBM: soyabean meal; Lys: lysine. Basal protein: mixture of animal and vegetable proteins of good balance providing 8% protein. Values with the same superscripts are not significantly different from each other (P<0.05).

Table 15. Comparative bodyweights and protein efficiency ratios of chickens fed normal maize and high-lysine maize [101].

Source of protein	Protein level (%)	Weight gain (g)	Feed/gain ratio	Source
Normal maize+safflower meal	20	173[c]	2.74[c]	[106]
Opaque-2+safflower meal	20	215[b]	2.25[b]	
Normal maize+soyabean meal	20	472[a]	1.46[a]	[106]
Normal maize+167.3 g/kg soyabean meal	15	300[c]	1.86[c]	
Opaque-2+I 67.3 g/kg soyabean meal	15	413[b]	1.64[b]	
Opaque-2+77.1 g/kg soyabean meal	15	293[c]	1.83[c]	
Opaque-2+214.5 g/kg soyabean meal	20	480[a]	1.38[a]	
Normal maize+289.6 g/kg soyabean meal	20	498[a]	1.40[a]	
QPM	9.7	708[b]	4.28[b]	[108]
Normal maize	9.7	532[c]	6.55[c]	
Normal maize+fishmeal	1.8	2017[a]	2.30[a]	
Normal maize+I 95 g/kg fishmeal	21	2149	2.60	[108]
QPM+195 g/kg fishmeal	21	2189	2.60	
QPM+175 g/kg fishmeal	20	2229	2.55	
QPM+155 g/kg fishmeal	19	2140	2.70	
QPM+135 g/kg fishmeal	18	2140	2.80	
Opaque-2+soyabean meal	22	390.2[c]	1.74[b]	[109, 110]
Opaque-2+soyabean meal	18	373.2[b]	1.83[c]	
Opaque-2+soyabean meal	14	320.5[a]	2.09[d]	
Normal maize+soyabean meal	22	411.5[d]	1.64[a]	
Normal maize+soyabean meal	18	372.8[b]	1.79[bc]	
Normal maize+soyabean meal	14	329.7[a]	2.09[d]	
NM+SBM	21	465	1.49	
OP+SBM	21	459	1.51	
NM+SBM	19	438	1.57	
OP+SBM	19	422	1.58	
NM+SBM	17	407[a]	1.67[a]	
OP+SBM	17	371[b]	1.72[b]	
NM+SBM	15	329	1.85[a]	
OP+SBM	15	332	1.81[b]	[104]
NM+SBM+Met	21	458	1.40	
OP+SBM+Met	21	468	1.39	
NM+SBM+Met	18	412	1.52	
OP+SBM+Met	18	428	1.48	
NM+SBM+Met	15	286[a]	1.84[a]	
OP+SBM+Met	15	369[b]	1.64[b]	

Source of protein	Protein level (%)	Weight gain (g)	Feed/gain ratio	Source
NM+SBM	20	451	1.45	
OP+SBM	20	463	1.42	
NM+SBM+amino acid	20	455	1.41	
NM+SBM	16	359[a]	1.76[a]	
OP+SBM	16	423[b]	1.58[b]	
NM+SBM+amino acid	16	412[b]	1.59[b]	
NM+SBM	12	163[a]	2.57[a]	
OP+SBM	12	243[b]	2.07[b]	
NM+SBM+amino acid	12	255[b]	2.09[b]	

NM: normal maize; SBM: Soyabean meal; OP: *Opaque*-2 maize; Met: methionine. Values with the same superscripts are not significantly different from each other (P<0.05). Not significant (P> 0.05).

Table 16. Comparative weight gains and feed/gain ratios of chickens fed normal maize and high-lysine maize at 3 weeks or 8 weeks of age.

Source of protein 2	Protein level (g/kg)	Weight gain (g)	Feed/gain ratio	Source
Normal maize+soyabean meal+methionine (0.2%)	200	481[a]	1.40[a]	[106]
Normal maize+safflower meal+methionine (0.19%)	200	158[c]	2.92[c]	
Normal maize+safflower meal+Lys+methionine (0.19%)	200	483[a]	1.53[a]	
Opaque-2+safflower mal+methionine (0.44%)	200	2I4[b]	2.11[b]	
Opaque-2+safflower meal+Lys+methionine (0.14%)	200	487[a]	1.52[a]	
Normal maize+safflower meal+Met (0.26%)	200	145[d]	2.55[a]	[106]
Opaque-2+safflower meal+Met (0.25%)	200	211[c]	2.15[a]	
Opaque-2 safflower meal+Met (0.26%)+Lys (0.15%)	200	364[a]	1.78[b]	
Normal maize+safflower meal+Met (0.25%)+Lys (0.15%)	200	270[b]	1.92[b]	
Normal maize+safflower meal+Met (0.20%)+Lys (0.14%)	150	163[c]	2.44[a]	
Opaque-2+safflower meal+met (0.13%)+Lys (0.18%)	ISO	282[b]	1.99[b]	
Opaque-2(0. 89 g/kg tryptophan)	95	129[a]	3.27[a]	[111]
Opaque-2+0.95 g/kg tryptophan	95	134[a]	3.09[a]	
Opaque-2+1.31 g/kg tryptophan	95	138[a]	2.99[a]	
Opaque-2+2.00 g/kg tryptophan	95	131[a]	3.09[a]	
Norma1 maize (0.59 g/kg tryptophan)	95	83[d]	4.24c	
Normal maize+0.95 g/kg tryptophan	95	115[bc]	3.54[a]	
Normal maize+1.31 g/kg tryptophan	95	127[a]	3.21[a]	
Normal maize+2.00 g/kg tryptophan	95	103[cd]	3.82[b]	
Opaque-2+sesame meal (0.97 g/kg tryptophan)	120	180*	2.43*	[111]
Opaque-2+sesame meal+1.20 g/kg tryptophan	120	162	2.62	

Source of protein 2	Protein level (g/kg)	Weight gain (g)	Feed/gain ratio	Source
Opaque-2+sesame meal+1.50 g/kg tryptophan	120	181	2.46	
Opaque-2+sesame meal+2.00 g/kg tryptophan	120	183	2.38	
Normal maize+sesame meal+(0.69 g/kg tryptophan)	120	146	2.84	
Normal maize+sesame meal+120 g/kg tryptophan	120	169	2.59	
Normal maize+sesame meal+1.50 g/kg tryptophan	120	166	2.63	
Normal maize+sesame meal+2.00 g/kg tryptophan	120	163	2.62	
Normal maize replacement with QPM	~220			[112]
Diet 1 (100:0)		634[b]	1.45[a]	
Diet 2 (75:25)		641[b]	1.38[b]	
Diet 3 (50:50)		675[a]	1.34[c]	
Diet 4 (25:75)		686[a]	1.35[c]	
Diet 5 (0:00)		673[a]	1.35[c]	
Diet 6 (100:lys)		676[a]	1.35[c]	

NM: normal maize; SBM: soyabean meal; OP: *opaque*-2 maize; Met: methionine. Values with the same superscripts are not significantly different from each other (P<0.05). *Not significant (P>0.05).

Table 17. Comparative weight gains and feed/gain ratios for chickens fed normal maize and high-lysine maize at 3 weeks of age.

In summary, the superior performance of chickens fed high-lysine maize has been attributed by all investigators to its higher lysine content and improved balance of essential amino acids.

Table 18 shows that when laying chickens were fed high-lysine maize at suboptimal protein levels, their egg production was significantly higher than that of their counterparts fed normal maize at the same suboptimal protein level. However, at optimal dietary protein levels, similar egg production was recorded. It was suggested that the difference in egg production at suboptimal dietary protein level was due to the higher lysine content of high-lysine maize [109, 110]. An advantage can thus be taken of the high lysine content of high-lysine maize in reducing dietary protein levels, that is, decreasing inclusion level of high protein feedstuffs like soyabean and fishmeal. A study in Ghana [113] found no significant difference in egg production when dietary protein level was reduced from 170 to 140 g/kg in diets in which QPM was incorporated (**Table 18**). Conflicting results have been obtained as to the effect of high-lysine maize on egg weight. Some studies [52, 109, 110, 113] found no significant difference in egg weight of hens fed either high-lysine or normal maize contrary to earlier results reported [106] that hens fed high-lysine maize produced heavier eggs than those fed normal maize. With respect to internal quality of eggs produced, no beneficial effect was reported when high-lysine maize was fed to hens instead of normal maize [52, 109, 110, 113]. Data presented in **Table 18** indicate that at suboptimal dietary protein levels, hens fed high-lysine maize diets utilised feed more efficiently than those fed normal maize diets even though the mean egg weights were similar.

Treatment	Protein level (g/kg)	Hen-day egg production (%)	Feed conversion (kg feed/kg eggs)	Egg weight (g)	Haugh units	Specific gravity	Source
QPM+SBM+Met	110	76.1[b]	1.48[ab]	53.8	86.0[c]	1.0826	[110]
QPM+SBM	140	80.6[bc]	1.43[b]	54.1	86.0[c]	1.0806	
QPM+SBM	170	84.2[c]	1.40[b]	55.5	86.7[c]	1.0815	
MN+SBM+Met	110	64.5[a]	1.60[a]	53.0	78.9[a]	1.0838	
NM+SBM	140	79.3[bc]	1.49[ab]	54.2	83.9[b]	1.0829	
NM+SBM	170	82.7[bc]	1.45[b]	54.8	79.8[a]	1.0838	
NM+SBM (90 g/kg)+Met	115.0	62.0[a]	1.91[a]	60.3	81.8	1.0760	[110]
QPM+SBM (90 g/kg)+Met	115.0	71.1[c]	1.65[c]	60.8	81.8	1.0748	
QPM+SBM (674 g/kg)+Met	107.6	71.1[c]	1.70[b]	62.0	82.3	1.0753	
QPM+SBM (45 g/kg)+Met	115.0	68.1 b	1.75[b]	59.9	83.6	1.0745	
NM+SBM	144	69.1	2.56[a]	64.1	–	–	[52]
HPC+SBM	158	70.7	2.35[b]	64.7	–	–	
NM+fishmeal (120 g/kg)	170	57.9	3.28	55.8	–	–	[113]
QPM+fishmeal (120 g/kg)	170	60.2	3.28	56.2	–	–	
QPM+fishmeal (100 g/kg)	160	55.6	3.59	56.4	–	–	
QPM+fisluneal (85 g/kg)	150	63.5	3.29	55.7	–	–	
QPM+fishmeal (65 g/kg)	140	60.2	3.36	55.5	–	–	
Normal maize 1	150	77.4[a]	2.06	58.9	87.1	–	[114]
QPM1	150	77.3[a]	2	58.5	89.6	–	
QPM2	140	72.0[ac]	2	58	89	–	
QPM3	140	64.3[b]	1.99	56.5	93.2	–	
Normal maize 2	140	66.8[b]	2.01	57.7	91.9	–	
Normal maize +0.07%Lysine	150	89.6[b]	2.15	59.2	97.1	–	[115]
QPM	149.4	91.0[a]	2.16	59.1	97.4	–	

Values with the same superscripts are not significantly different from each other (P<0.05).

Table 18. Comparative performance of laying chickens fed normal maize or high-lysine maize.

9. Future challenges of maize utilisation in poultry diets

The major future challenges confronting maize utilisation in poultry production include the following:

- Adverse effects of climate change on maize production have been reported in tropical and subtropical regions. These include frequent droughts, heat, increased temperature and inadequate rainfall during the growing season and water-logging. It has been estimated that one quarter of the global maize areas is affected by drought in any given year [116].

- Competition between humans and animal agriculture. Maize is increasingly being used for human food and other industrial purposes including biofuel production. Thus, in a world where the global population is continually increasing, the argument that producing feed for livestock conflicts with feeding hungry people is likely to continue for some years [117].

- Some challenges for widespread adoption of QPM in developing countries have been described [118], which include lack of profitable markets for commercial producers and lack of government incentive to encourage adoption by subsidising the price of QPM seed.

10. Conclusion

Normal maize is the most widely used single grain in poultry feeding due to a combination of desirable nutritional characteristics such as high energy, low fibre and easy digestibility. However, its low protein content and deficiencies of the protein in lysine and tryptophan have been a major nutritional concern for feeding poultry. Therefore, improvement in the protein quality of normal maize through the development of a new class of *opaque-2* maize known as "quality protein maize (QPM)" has been a major boost for poultry production, particularly in developing countries, where dietary supplemental protein is either expensive or imported. The benefits of feeding QPM grain to poultry are greater weight gain and more efficient feed conversion as well as less supplemental protein cost.

There are, however, pertinent challenges confronting maize production such as adverse impact of climate change, stiff competition between humans and animal agriculture, and some challenges of widespread adoption of QPM. There is a need to overcome these and other challenges in order to increase cheap meat production to meet the needs of the growing global population.

Author details

Herbert K. Dei

Address all correspondence to: hkdei@yahoo.com

Department of Animal Science, Faculty of Agriculture, University for Development Studies, Tamale, Ghana

References

[1] OECD/FAO, Organisation for Economic Co-operation and Development/Food and Agriculture Organisation. 2014. OECD-FAO Agricultural Outlook 2014–2023. OECD/FAO Agricultural Outlook 2014, OECD Publishing, Paris.

[2] Larbier M, Leclercq B. Nutrition and Feeding of Poultry (J. Wiseman, Trans. & Ed.). Loughborough: Nottingham University Press; 1994. 305 p.

[3] Summers JD. Factors influencing food intake in practice—Broilers. In: 7th Nutrition Conference for Feed Manufactures (H. Swan and D. Lewis, Eds.); London. 1974. p. 127–140.

[4] Black JL. Variation in nutritional value of cereal grains across livestock species. Proceedings of Australian Poultry Science Symposium. 2001; 13: 22–29.

[5] Stamen B. Feed preference index on cereal grains for poultry. 2010. Available from: http://hdl.handle.net/1811/45500 [Accessed 2016-06-14].

[6] NRC, National Research Council. Quality Protein Maize. Washington: National Academy Press; 1988. 83 p.

[7] FAO, Food and Agriculture Organisation. Statistical Yearbook of FAO. Rome: FAO Publications; 2015.

[8] USDA, United States Department of Agriculture. World Agriculture Supply and Demand Estimates. Washington: Economic Research Services; 2016.

[9] Potter NN, Hotchkiss IH. Food Science, 5th edn. New York: Chapman and Hall; 1995. p. 381–394.

[10] FAO, Food and Agriculture Organisation. Statistical Yearbook of FAO. Rome: FAO Publications; 2005.

[11] Watson SA. Corn marketing, processing and utilisation. In: Sprague GF, Dudley JW, editors. Corn and corn improvement. Agronomy Journal. 1988; 18: 881–940.

[12] Castanon F, Han Y, Parsons CM. Protein quality and metabolisable energy of corn gluten feed. Poultry Science. 1990; 69: 1165–1173.

[13] Kling IG. Quality and Nutrition of Maize; IITA Research Guide 33. Ibadan: International Institute of Tropical Agriculture. 1991; p. 9–21.

[14] Watson SA, Ramstad PR. Corn Chemistry and Technology. St Paul: American Association of Cereal Chemists. 1987; p.605.

[15] IITA, International Institute of Tropical Agriculture. Maize crop. Maize Production Manual. 1982; 1(8), 1–2.

[16] MacDonald P, Edwards RA, Greenhalgh J F. Animal Nutrition, 4th edn. Essex: Longman Group Ltd; 1988. 479 p.

[17] Lawrence TL. Feeding value of cereals and concentrates. In: Orskov ER, editor. World Animal Science. Amsterdam: Elsevier Science Publishers BV; 1988. p. 129–146.

[18] Fetuga BL, Babatunde GM, Oyenuga VA. Comparison of the energy values of some feed ingredients for the chick, rat and pig. Journal of Agricultural Science. 1979; 17: 3–11.

[19] MAFF, Ministry of Agriculture, Fisheries and Food. Tables of feed composition and energy allowances for ruminants. Middlesex: Pinner; 1975.

[20] Eggum BO. The nutritive quality of cereals. Cereal Research. 1977; 5: 153–157.

[21] FAO, Food and Agriculture Organisation. Improvement of nutritional quality of food crops. FAO Plant Production and Protection Paper 34. Rome: FAO; 1981.

[22] Shurpalekar KS, Sundaravalli OE, Rao MN. *In vitro* and *in vivo* digestibility of legume carbohydrates. Nutrition Report International. 1979; 19: 111–117.

[23] Martinez BF, Sevilla PE, Bjarnason M. Wet milling comparison of quality protein maize and normal maize. Journal of the Science of Food and Agriculture. 1996; 71: 156–162.

[24] Morrison WR. Cereal lipids. Proceedings of Nutrition Society. 1977; 36: 143–148.

[25] West CE, Pepping F, Temalilwa CR. The Composition of Foods Commonly Eaten in East Africa. Wageningen: Centre for Tropical Agriculture (CTA); 1988. p. 84.

[26] Card LE, Nesheim MC. Poultry Production, 11th edn. Philadelphia: Lea Fabiger; 1972. p. 210–243.

[27] Blair JC, Harper CD, McNab JM, Mitchell GG, Scougall RK. Analytical Data of Poultry Feedstuffs. 1: Genral Amino Acid Analysis, 1977–1980. Occasional Publication, No.1. Edinburgh: ARC-Poultry Research Centre; 1981.

[28] Johnson VA, Lay CL. Genetic improvement of plant proteins. Journal of Agriculture and Food Chemistry. 1974; 22: 558–566.

[29] Kies C, Fox HM. Interrelationships of leucine with lysine, tryptophan, niacin as they influence protein value of cereal grains for humans. Cereal Chemistry. 1972; 49: 223–231.

[30] Bressani R. Protein quality of high-lysine maize for humans. Cereal Food World. 1991; 36: 806–811.

[31] Bressani R. Improving maize diets with amino acid and protein supplements. In: High-quality protein maize. CMMYT-Purdue International Symposium on Protein Quality Maize. Hutchinson and Ross Inc., Stroudsburg. Pennsylvania: CMMYT; 1975. p. 38–57.

[32] Bressani R, Marenco E. The enrichment of lime-treated corn flour with proteins, lysine, tryptophan and vitamins. Journal of Agriculture and Food Chemistry. 1963; 6: 517–522.

[33] Bressani R, Elias LG. All vegetable mixtures for human feeding. The development of INCAP vegetable mixture based on soyabean flour. Journal of Food Science. 1967; 31: 626–631.

[34] Bressani R. The importance of maize for human nutrition in Latin America and other countries. In: Bressani R, Braham JE, Behar M, editors. Nutritional improvement of maize. Guatamala: INCAP; 1972. p. 5–29.

[35] Bressani R, Elias EG. Processed vegetable protein mixtures for human consumption in developing countries. Advanced Food Research. 1968; 16: 1–13.

[36] Bressani R. Amino acid supplementation of cereal grain flours tested in children. In: Scrimshaw NS, Altschul AM, editors. Amino Acid Fortification of Protein Foods. Cambridge: MIT Press; 1971. p. 184–204.

[37] Bressani R, Elias LG, Viteri F. Complementation effects in rats fed combinations of soyabean flour and whole maize flour at a constant level of dietary protein. Journal of Food Science. 1974; 39: 577–580.

[38] Osborne TB, Mendel LB. Amino acids in nutrition and growth. Journal of Biology and Chemistry. 1914; 17: 325–349.

[39] Gallo JT, Maner JH, Jimenez I. Nutritive value of opaque-2 corn for growing pigs. Journal of Animal Science. 1968; 28: 1152 (Abstr.).

[40] Baker DH, Becker DE, Jensen AH, Harmon BG. Protein source and level for pregnant gilts: a comparison of corn, opaque-2 corn and corn-soyabean diets. Journal of Animal Science. 1970; 30: 364–369.

[41] Viteri FB, Martinez C, Bressani R. Evaluation of the protein quality of common maize, *opaque-2* maize supplementation with amino acids and other sources of protein. In: Bressani R, Brahman JE, Behar MM, editors. Nutritional improvement of maize. Guatemala: NCAP; 1972.

[42] Boomgaart J, Baker DH. The lysine requirements of the chick at five protein levels. Poultry Science. 1970; 49: 1369 (Abstr.).

[43] Howe EE, Jansen GR, Gilfillan EW. Amino acid supplementation of cereal grains as related to the world food supply. American Journal of Clinical Nutrition. 1965; 16: 315–321.

[44] Leonard WH, Martin IH. Cereal Crops. New York: Macmillan Co.; 1963. p. 131–360.

[45] FAO, Food and Agriculture Organisation. 1992. Maize in human nutrition. FAO Report Series 25, Rome, Italy.

[46] Bressani R. Protein complementation of foods. In: Karmas E, Harris RS, editors.

Nutritional Evaluation of Food Processing, 3rd edn. New York: Van Nostrand Reinhold Company; 1988. p. 627–657.

[47] Duffus CM, Slaughter JC. Seeds and Their Uses. New York: John Wiley and Sons Ltd; 1980. p. 95.

[48] Maner IH. Quality Protein Maize in Swine Feeding and Nutrition. Arlington: International Agriculture Development Services; 1983.

[49] Earley EB, De Turk EE. Corn protein and soil fertility. In: What's New in the Production, Storage and Utilisation of Hybrid Seed Corn. Chicago: America Seed Traders Association; Hybrid Corn Division Annual Report. 1948; 3: 84–95.

[50] Pollmer WG, Eberhard D, Klein D, Dhillon BS. Studies on maize hybrids involving inbred lines with varying protein content. Pflanzenzuchtag. 1978; 80: 142–148.

[51] Boyat M, Deriux M, Kaan F, Rautou S. Maize breeding for improvement of kernel protein content using Illinois high protein population. In: Pollmer WG, editor. Improvement of Quality Traits of Maize for Grain and Silage. Pflanzenzuchtag. 1980; 80: 173–183.

[52] Bond PL, Sullivan TW, Douglas JH, Robeson LG, Baier JG. Composition and nutritive value of experimental high-protein corn in the diets of broiler and laying hens. Poultry Science. 1991; 70: 1578–1574.

[53] Sprague GF. Corn and Corn Improvement. New York: Academic Press Inc.; 1955. p. 613–635.

[54] Leng ER, Curtis JJ, Shekleton MC. Niacin content of waxy, sugary and dent F segregating kernels in corn. Science. 1950; 111: 665–666.

[55] Dudley JW, Lambert RJ. Ninety generations of selection for oil and protein in maize. Maydica. 1992; 37: 81–87.

[56] Woodworth CM, Leng ER, Jugenheimer RW. Fifty generations of selection for protein and oil in corn. Agronomy Journal. 1952; 44: 60–64.

[57] Hancock 1D, Peo Jr. ER, Lewis AI, Kniep KR, Mason SC. Effects of irrigation and nitrogen utilisation of normal and high lysine corn on protein utilisation by the growing rat. Nutrition Report International. 1988; 38: 413–422.

[58] Douglas JH, Sullivan TW, Bond PL, Struwe FJ. Nutrient composition and metabolisable energy value of selected grain sorghum varieties and yellow corn in broiler diets. Poultry Science. 1990; 69: 1147–1155.

[59] Nelson TS, May MA, Miles Jr. RD. Digestible and Metabolisable energy content of feed ingredient for rats. Journal of Animal Science. 1974; 38: 555–558.

[60] Scott HM, Matterson LD, Singsen EP. Nutritional factors influencing growth and efficiency of feed utilisation. 1. The effect of the source of carbohydrate. Poultry Science. 1947; 26: 554–559.

[61] Hillman RI, Kienholz EW, Shroder CD. The effect of corn gluten feed in chicken and turkey breeder diets. Poultry Science. 1973; 52: 2309 (Abstr.).

[62] Jones RW. Corn co-products as feed ingredients for swine: effects on growth, carcass composition, fibre and amino acid digestibility [PhD thesis]. Urbana-Champaign: University of Illinois; 1987.

[63] Owings WI, Sell IL, Ferket P, Hasiak R1. Growth performance and carcass composition of turkey hens fed corn gluten feed. Poultry Science. 1988; 67: 585–589.

[64] Corn Refiners Association. Corn Wet-milled Feed Products, 2nd edn. Washington: Corn Refiners Association; 1982.

[65] Castanon F. Nutritional evaluation of corn gluten feed by poultry [MSc thesis]. Urbana-Champaign: University of Illinois; 1988.

[66] Sibbald 1R. The T.M.E. System of Feed Evaluation: Methodology, Feed Composition Data and Bibliography. Ottawa: Animal Research Centre Contribution 85–19; 1986.

[67] Ellis NR, Kauffman WR, Miller CO. Composition of principal feedstuffs used for livestock. In: Food for Life. Washington: USDA Yearbook of Agriculture; 1939. p. 1065–1074.

[68] Forster JF, Yang JT, Yui NH. Extraction and electrophoretic analysis of the proteins of corn. Cereal Chemistry. 1950; 27: 477.

[69] Bressani R, Mertz ET. Studies on corn proteins. IV. Protein and amino acid content of different corn varieties. Cereal Chemistry. 1958; 35: 227.

[70] Rathman DR. Zein, An Annotated Bibliography (1891–1953). Pittsburg: Mellon Institute; 1954.

[71] Mosse J. Alcohol-soluble proteins of cereal grains. Feedstuff Proceedings. 1966; 25: 1663–1669.

[72] Bressani R, Valiente AT, Tejada C. All vegetable protein mixtures for human feeding. VI. The value of combinations of lime-treated corn and cooked black beans. Journal of Food Science. 1962; 27: 394.

[73] Tello F, Alvarez-Tostado MA, Alvarado G. A study on the improvement of essential amino acid balance of corn proteins. 1. Correlation between racial and varietal characteristics and lysine levels of corn. Cereal Chemistry. 1965; 42: 368–372.

[74] Mertz ET, Bates LS, Nelson OE. Mutant gene that changes protein composition and increases lysine content of maize endosperm. Science. 1964; 145: 279–280.

[75] Bates LS. Amino acid analysis. In: Proceedings of High-Lysine Corn Conference. Washington: Corn Refiners Association; 1966. p. 61–66.

[76] Ahuja VP, Singh J, Naik MS. Amino acid imbalance of proteins of maize and sorghum. Indian Journal of Genetics. 1970; 30:727–731.

[77] Nelson OE, Mertz ET, Bates LS. Second mutant gene affecting the amino acid pattern of maize endosperm protein. Science. 1965; 150: 1469–1470.

[78] Misra PS, Jambunatha R, Mertz ET, Glover DV, Barosa HM, McWhirter KS. Endosperm synthesis in maize mutants with increased lysine content. Science. 1972; 176: 1425–1427.

[79] Ma Y, Nelson OE. Amino acid composition and storage proteins in two new high-lysine mutants in maize. Cereal Chemistry. 1975; 52: 412–418.

[80] Paez AV, Helm IL, Zuber MS. Kernel opacity development and moisture content within maize ears segregating for *opaque-2* and floury-2. Agronomy Journal. 1969; 61: 443–445.

[81] Zarkadas CG, Yu Z, Hamilton RI, Pattison PL, Rose NGW. Comparison between the protein quality of northern adapted cultivars of common maize and quality protein maize. Journal of Agricultural and Food Chemistry. 1995; 43: 84–93.

[82] Lambert RI, Alexander DE, Dudley IW. Relative performance of normal and modified protein (*opaque-2*) maize hybrids. Crop Science. 1969; 9: 242–243.

[83] Sreeramulu C, Bauman LF. Yield components and protein quality of *opaque-2* and normal diallels of maize. Crop Science. 1970; 10: 262–265.

[84] Dilmer RJ, Report on kernel structure and wet milling of high lysine corn. In: Proceedings of High-Lysine Corn Conference. Washington: Corn Refiners Association; 1966. p. 121–127.

[85] Gupta SC, Asnani VL Khare BP. Effects of *opaque-2* gene in maize on the extent of infestation by Sitophilus oryzae. Journal of Stored Products Research. 1970; 6: 191–194.

[86] Chartterjee SM, Panwar VPS, Siddiqui KH, Young WR, Marwala KK. Field screening of some promising germplasms against Chilo zonellus Swinhoe under artificial infestation. Indian Journal of Entomology. 1970; 32: 167–170.

[87] Asnani VL, Gupta SC. Effect of incorporation of opaque-2 gene on yield and yield components in four composites of maize. Indian Journal of Genetics. 1970; 30:377–382.

[88] Vasal SK, Villegas E, Bjarnason M, Gelaw B, Goertz P. Genetic modifiers and breeding strategies in developing hard-endosperm *opaque-2* materials. In: Pollmer WG, Phipps RH, editors. Improvement of Quality Traits of Maize for Grain and Silage Use. London: Martinus Nijhoff; 1980. p. 37–73.

[89] Ortega EI, Bates LS. Biochemical and Agronomic studies of two modified hard-endosperm *opaque-Z* maize populations. Cereal Chemistry. 1983; 60:107–111.

[90] Bjarnason M, Vasal SK. Breeding of quality protein maize (QPM). In: Janick J, editor. Plant Breeding Review 9. New York: Wiley; 1992. p. 181–216.

[91] Twumasi-Afriyie S, Badu-Apraku B, Sallah PYK. 1992. The potential of maize as a source of quality protein for Ghana. In: Proceeding of the 12th National Maize and Legume workshop. 25–27 March 1992; Kumasi.

[92] Liu ZX, Jie SF, Guo XF, Xu IF, Wang LM. Research on selected high quality protein maize hybrids. Maize Genetics Newsletter (March, 1993). 1993; 67:50–52.

[93] CIMMYT, Centro International de Mejoramiento de Maiz Y Trigo. Boosting protein quality in maize. CIMMYT Research Highlights. Mexico City: CIMMYT; 1985.

[94] Knabe DA, Sullivan IS, Burgoon KG. QPM as a swine feed. In: Mertz ET, editor. Quality Protein Maize. Washington: The American Association of Cereal Chemists; 1992. p. 225–239.

[95] Villegas EM, Eggum BO, Kohli MM, Vasal SK. Progress in nutritional improvements of maize and triticale. Food Nutrition Bulletin. 1980; 2:17–24.

[96] Ortega E1, Villegas E, Bjarnason M, Short K. Changes in dry matter and protein fractions during kernel development of quality protein maize. Cereal Chemistry. 1991; 68:482–486.

[97] Kniep KR, Mason SC. Lysine and protein content of normal and *opaque-2* maize grain as influenced by irrigation and nitrogen. Crop Science. 1991; 31:177–181.

[98] Gomez, M.H., S.O. Serna-Saldivar, I.Y. Corujo, A.I. Bockholt and L.W. Rooney. Wet milling properties of quality protein maize and regular corns. In: Mertz ET, editor. Quality Protein Maize. Washington: The American Association of Cereal Chemists; 1992. p. 239–260.

[99] Ahenkorah K, Twumasi-Afriyie S, Nagai D, Haag W, Asirifi-Yeboah K, Dzah BD. QPM (Obatanpa) as food source in Ghana: nutritional composition, rat growth and protein quality trials. In: Proceeding of the Ghana Animal Science Association Symposium. 1994; 22:25–30.

[100] Bressani R, Elias LG, Gomez-Brenes RA. Protein quality of opaque-2 corn evaluation in rats. Journal of Nutrition. 1969; 97:173–180.

[101] Mertz ET, Veron OA, Bates LS, Nelson O.E. Growth of rats fed on *opaque-2* maize. Science. 1965; 148:1741–1742.

[102] Rivera PH, Peo Jr. ER, Flowerday D, Crenshaw TD, Moger BD, Cunningham PR. Effect of maturity and drying temperature on nutritional quality and amino acid availability of normal and *opaque-2* corn for rats and swine. Journal of Animal Science. 1978; 46:1024–1085.

[103] Bosque CM de, Bressani R. Effect of protein quality of corn on liver retinol reserves, food intake, weight gain and feed efficiency in rats. In: Quality Protein Maize. Washington: National Academy Press; 1987. p. 49.

[104] Cromwell GL, Rogler JC, Featherston WR, Pickett RA. Nutritional value of opaque-2 corn for the broiler chick. Poultry Science. 1967; 46:704–712.

[105] Cromwell GL, Rogler JC, Featherston WR, Pickett RA. A comparison of the nutritive value of opaque-2, floury-2 and normal corn for the chick. Poultry Science. 1968; 47:840–847.

[106] Fonseca JB, Featherston WR, Rogler JC, Cline TR. A comparison of the nutritive value of opaque-2 and normal corn for the laying hen. Poultry Science. 1970; 49:532–537.

[107] Fernandez R, Lucas E, Ginnis JC. Comparative nutritional value of different cereal grains as protein sources in a modified chic bioassay. Poultry Science. 1974; 53:39–46.

[108] Osei SA, Atuahene CC, Donkoh A, Kwateng K, Ahenkorah K, Dzah BD, Haag W, Twumasi-Afriyie S. Further studies on the use of QPM as a feed ingredient for broiler chickens. In: Proceedings of the Ghana Animal Science Association Symposium. 1994; 22:51–55.

[109] Chi MS, Speers GM. Nutritional value of high-lysine corn for broiler chick. Poultry Science. 1973; 52:1148–1157.

[110] Chi MS, Speers GM. A comparison of the nutritional value of high-lysine, floury-2 corn and normal corn for the laying hen. Poultry Science. 1973; 52:1138–1147.

[111] Cuca GM, Pró MA. Tryptophan and Methionine supplementation of opque-2 and normal corn for chicks. Poultry Science. 1972; 51:787–791.

[112] Panda AK, Raju MVLN, Rama Rao SV, Lavanya G, Pradeep Kumar Reddy E, Shyam Sunder G. Replacement of normal maize with quality protein maize on performance, immune response and carcass characteristics of broiler chickens. Asia-Australian Journal of Animal Science. 2010; 23:1626–1631.

[113] Osei SA, Donkoh A, Atuahene CC, Hagan 1A, Tuah AK. Quality protein maize in the diet of laying chickens. In: Abstracts and Students' Papers Contributed In International Youth Programme. Vol.IV, World's Poultry Congress (2–5 September 1996); New Delhi: 1996.

[114] Osei SA, Dei HK, Tuah AK. Evaluation of quality protein maize as a feed ingredient for layer pullet. Journal of Animal Feed Science. 1999; 8:181–189.

[115] Zhai S. Nutritional evaluation and utilisation of quality protein maize (Zhong Dan 9409) in laying hen feed [thesis]. Shaanxi: Northwestern Agricultural and Forestry University of Science and Technology; 2002.

[116] Heisey PW, Edmeades GO. Maize production in drought-stressed environments: technical options and research resource allocation. In: Part 1 of CIMMYT 1997/1998 World Facts and Trends. Mexico: CIMMYT; 1999.

[117] CAST, Council for Agricultural Science and Technology. Animal Feed vs. Human Food: Challenges and Opportunities in Sustaining Animal Agriculture Toward 2050. Issue Paper 53. Ames: CAST; 2013.

The Effect of Age on Growth Performance and Carcass Quality Parameters in Different Poultry Species

Daria Murawska

Abstract

In recent years, a steady increase in global poultry meat production has been witnessed, accompanied by an increase in a major portion of a poultry carcass, referred to as the inedible portion. In poultry, edible components include meat, skin with subcutaneous fat and giblets (gizzard, liver, and heart) and sometimes also abdominal fat in waterfowl. Age, together with species and environmental conditions, is one of the key factors affecting body growth rate. In four poultry species, chickens, turkeys, Pekin ducks, and geese, an increase in body weight is accompanied by an increase in edible weight and a decrease in inedible weight in the carcass, and more significant age-related changes occur in turkeys and broiler chickens than in ducks and geese. The highest increase in the content of edible components expressed as a percentage of total body weight is noted in turkeys (20% in males, 25% in females), followed by broiler chickens (19.4%), ducks (17.1%), and geese (only 8.2%). Gallinaceous birds have also a higher content of muscle tissue and a lower content of skin (including subcutaneous fat) and bones than waterfowl.

Keywords: tissue components, edible components, non-edible components, poultry species

1. Introduction

Age, together with species and environmental conditions, belongs to the key factors affecting the growth rate of birds. Age has a significant effect also on carcass tissue composition, and the most profound changes, both qualitative and quantitative, occur at early life stages which are relatively short (compared with the lifespan) and characterized by rapid growth.

Selection progress in meat-type poultry (particularly in broiler chickens and turkeys) has contributed to an increase in their body weight, improved carcass composition, a shorter production period, and a substantial rise in carcass dressing percentage, which results from an increase in the content of edible portions in the total body weight of birds, accompanied by a decrease in the content of inedible components treated as slaughter offal [1]. The total percentage content of edible and inedible components in the carcasses of different poultry species is an important economic consideration since the waste load from meat processing plants has to be effectively managed or disposed of [2–7].

In poultry, edible components include meat, skin with subcutaneous fat, and giblets (gizzard, liver, and heart), sometimes also abdominal fat in waterfowl. Fast-growing birds need concentrated feeds in an adequate amount and appropriate form. Changes in the nutritional regime affect the function of gastrointestinal tract segments and the weight of internal organs [8–10]. The growth rate of muscle tissue—the most valuable edible component—is faster than the growth rates of internal organs and other body parts classified as inedible, such as feathers, blood, head, and others [11–16].

In growing birds, significant age-related changes can be observed not only in carcass tissue composition but also in the location of lean meat, skin with a fat layer, and bones in the carcass [11, 17–20]. Carcass quality is also determined by the distribution of tissue components. Lean meat should be located in the most valuable carcass parts—the breast and legs, rather than in the neck, wings, and back, which are less valuable. The tissue composition of poultry carcasses changes with age because the growth rates of tissue components vary across species.

Carcass parts and tissue components can be classified as edible and nonedible. Methodological differences in carcass preparation (carcass with or without the neck and wing tips) and the classification of abdominal fat as an edible or a nonedible component make it difficult to compare the research findings reported by various authors. The present report summarizes mostly the results of own studies due to the uniform methodological approach to carcass processing and classification of edible and nonedible carcass components in the analyzed poultry species, that is, Ross 308 broiler chickens [13], Pekin ducks [14], BIG 6 turkeys [15, 20], and Koluda® White geese [16, 21].

2. The effect of age on the growth rates of tissues and organs in broiler chickens, turkeys, ducks, and geese

Selective breeding carried out over many generations has resulted in specialized poultry meat and egg production based upon specialized meat stock and egg stock, including single-purpose strains and lines selected separately. Breeds and strains of meat-type poultry are intensely selected for fast growth rate, high body weight, high carcass dressing percentage, and lean tissue deposition. Dual-purpose birds, raised for both meat and eggs, are less cost-effective than specialized layers and broilers in large-scale commercial poultry farming.

2.1. Body weight and carcass weight

Body weight is highly heritable and easy to measure [22, 23]. Body weight variation over time considerably affects production effectiveness in meat-type poultry. The growth rates of birds vary with age and across species. In comparison with ducks and geese, chickens and turkeys grow slower in the first weeks of life. The growth rates of ducks and geese decrease at 7 and 10 weeks of age, respectively. Migratory birds such as ducks and geese have to be capable of flight within a short time after hatching, so they gain weight rapidly during early life stages. Despite domestication and long-term selection aimed at eliminating atavistic features, adjustments enhancing flight capabilities can still be observed in modern ducks and geese [24, 25]. Broiler chickens have been raised since the 1920s, but initially only surplus cockerels from the laying stock were used as a source of poultry meat. In 1923, the average body weight of meat-type cockerels at 16 weeks of age was 1 kg [26]. Today, the average slaughter age of broiler chickens is 35–42 days at a body weight of 2.10–2.80 kg [27]. Between 1 and 10 weeks of age, the body weight and carcass weight of broiler chickens increase nearly 30-fold and over 40-fold, respectively [13] (**Figure 1a**). If the rearing period is longer than 35–42 days, the produced broilers have higher body weight and muscle yield, but also higher carcass fat content.

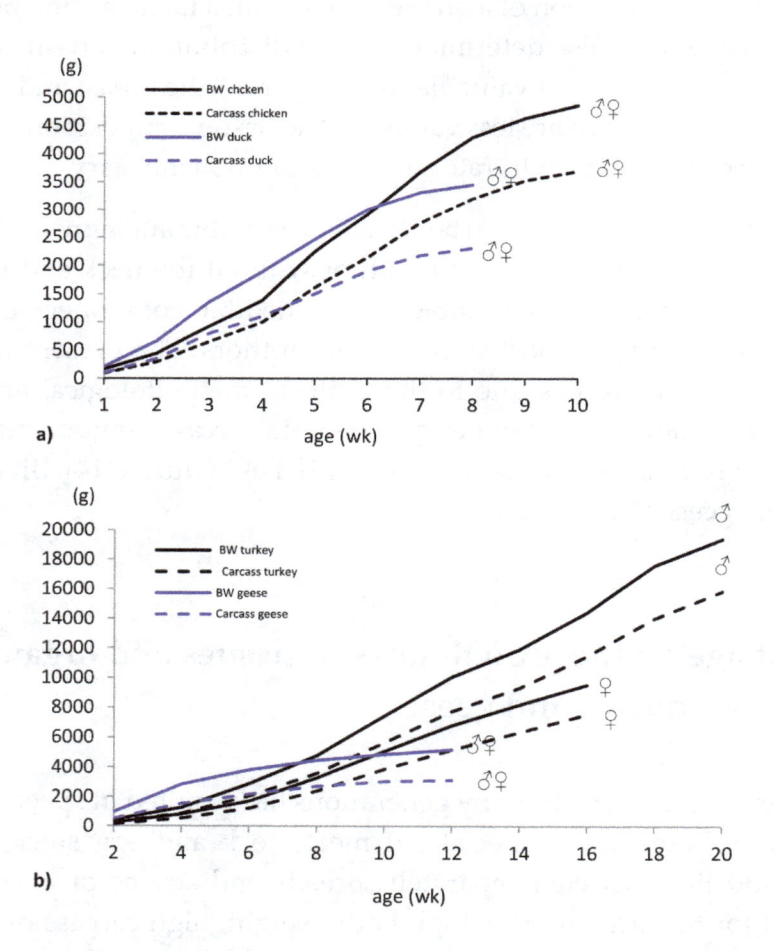

Figure 1. Arithmetic means (x) for the body weight and carcass weight of (a) chickens and ducks, (b) turkeys and geese (g).

Turkeys are the second most popular meat-type poultry species, after broiler chickens. Turkeys are heaviest among the four most common poultry species raised for meat. In comparison with other poultry species, turkeys are characterized by the highest carcass dressing percentage and high feed efficiency [4–6]. Turkey carcasses have high lean content and relatively low fat content [28]. Based on their body weight, modern commercial hybrid turkeys can be divided into heavy type, medium-heavy type, and medium type. Depending on the type, the body weight of males and females at 20 weeks of age ranges from 16.8 to 21 kg and from 9.3 to 11 kg, respectively. Due to considerable sexual dimorphism in the body weight of turkeys, males are reared over a longer period of time than females. Turkeys are characterized by the highest carcass dressing percentage among all poultry species, which reaches 81% in 16-week-old females and 84% in 21-week-old males [29].

The average body weight of male and female BIG 6 turkeys increases 45.6-fold between 2 and 20 weeks of age, and 40.8-fold between 2 and 16 weeks of age, respectively. The respective values for carcass weight are 58.3-fold (♂) and 57.7-fold (♀). A nearly 58-fold increase in carcass weight is observed already at 16 weeks of age in females and at 20 weeks of age in males, as compared with initial carcass weight at 2 weeks of age [15, 20] (**Figure 1b**).

Ducks can be divided into meat type, egg-layer type, and dual purpose. Broiler ducks are raised to 7–8 weeks of age and slaughtered at a body weight of approximately 3.5 kg. Hybrid Pekin ducks are used commercially for meat production [9]. Between 1 and 8 weeks of age, the average body weight and carcass weight of Pekin ducks increase 16.5-fold and 23.3-fold, respectively [14] (**Figure 1a**).

Geese are also raised mostly for meat, whereas egg, feather, and down production remains marginal. Geese are slaughtered at different ages, depending on breed and management system [30, 31]. Broiler-type geese are raised intensively to 10–12 weeks of age, and they are slaughtered at a body weight of approximately 5.2 kg. Geese have a fast initial growth rate. From 1 day to 12 weeks of age, their body weight increases over 34-fold [21], compared with a 5.3-fold increase in body weight and a 6.2-fold increase in carcass weight between 2 and 12 weeks of age [16] (**Figure 1b**).

Broiler chickens intended for grilling, with carcass weight of approximately 1 kg, can be slaughtered at 28 days of age. Medium-heavy female turkeys that are to be roasted whole can be slaughtered at 12 weeks of age. However, market demand for such products tends to be seasonal and remains low. Large carcasses that can be divided into retail and market-ready cuts or used for further processing are preferred, and this applies also to ducks and geese.

2.2. Growth rates of tissue components

The whole-body growth rate of birds varies over time, similarly as the growth rates of tissue components, which affects the proportions between carcass parts and tissue components [11, 17–19]. Chicks, poults, ducklings, and goslings are characterized by relatively slow growth of muscle tissue and adipose tissue, and fast growth of bones and internal organs. The content of muscle tissue and skin (including subcutaneous fat) increases significantly with age, whereas the content of giblets (liver, heart, gizzard), bone, and slaughter offal (excluding abdominal fat and feathers) decreases [13–16].

2.2.1. Lean meat

In birds raised specifically for meat production, both the percentage and the distribution of lean tissue in the carcass are important consideration. In poultry, breast and leg muscles constitute the most valuable carcass portion. The growth rates of muscle groups vary across poultry species. Similar age-related changes in muscle growth can be observed in chickens and turkeys, and in ducks and geese.

In broiler chickens, carcass lean content increases approximately 49-fold between 1 and 10 weeks of age. Over this period, the percentage of muscle tissue increases from 30.9 to 51.3% of total body weight, by 16.5% until week 6, and by approximately 4% between 6 and 10 weeks of age [13] (**Figure 2a**). In the carcasses of 2-week-old broiler chickens, approximately 36 and 35% of lean meat is located in the breast and legs, respectively. As birds grow older, the percentage of lean tissue increases in the breast (to 44%) and decreases insignificantly in the legs (to 32%), relative to the total lean content of the carcass [17].

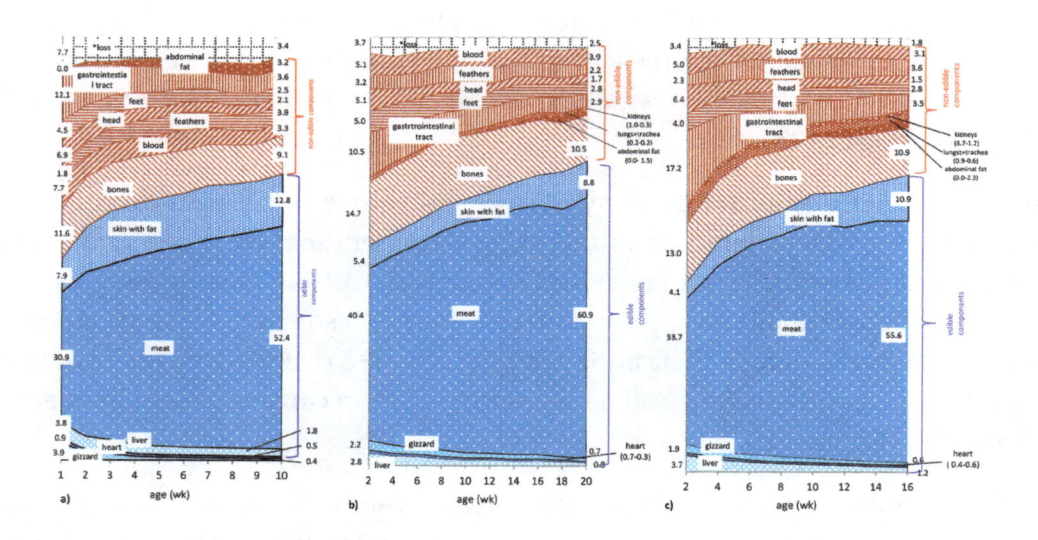

Figure 2. Percentage content of particular components in the body weight of (a) broiler chickens and turkeys, (b) males, (c) females. *loss = body weight loss during post-slaughter processing and dissection (%) [13, 15].

Turkey carcasses have high lean content [5, 6]. The growth rates of muscle tissue vary between carcass parts. In male BIG 6 turkeys, the weight of breast muscles increases approximately 77.1-fold between 2 and 20 weeks of age, the weight of leg muscles increases 75.4-fold, and the combined weight of the remaining muscles is only 54.4-fold. In female BIG 6 turkeys, the weights of breast muscles and leg muscles increase 66.7-fold and 70.5-fold, respectively, between 2 and 16 weeks of age, whereas the combined weight of the remaining muscles increases 60.5-fold [15, 20]. Muscle tissue accounts for approximately 40 and 34% of the total body weight in 2-week-old males and females, respectively, and increases to approximately 61 and 56%, respectively, at slaughter, that is, at 20 weeks of age in males and 16 weeks in females

[15]. Changes in the distribution of lean meat in the carcass can be observed in growing turkeys. In males, lean meat content (relative to total lean weight in the carcass) increases in the breast (by 3.4%) and legs (by 1.8%), and decreases in the wings (by 2.8%), back (by 1.5%), and neck (by 1%). In females, muscle tissue distribution in the carcass is similar, with a higher (by 4.2%) percentage of lean meat located in the breast. The most valuable muscles, that is, breast and leg muscles, account for 72 and 74% of total meat weight in males and females at slaughter age, respectively [20].

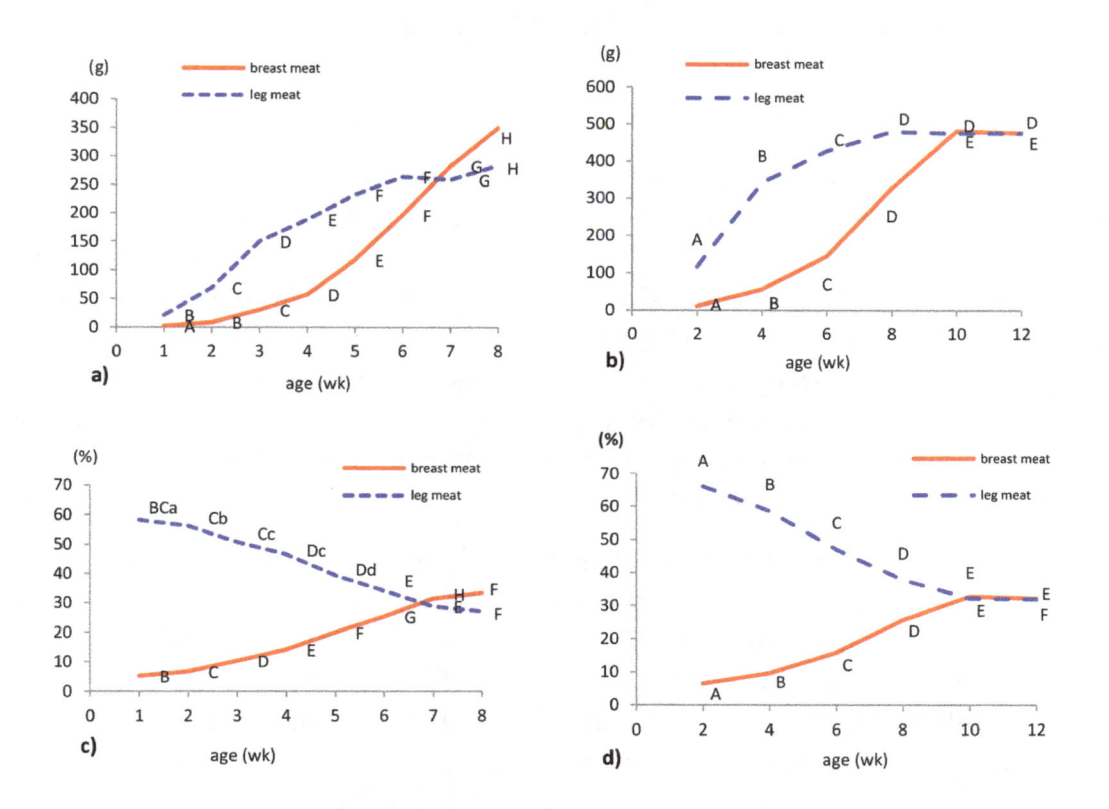

Figure 3. Arithmetic means ($\bar{X}$) for weight of breast muscles and leg muscles (g): (a) ducks, (b) geese and percentage share of breast muscles and leg muscles in total lean weight, subject to the age of birds (%), (c) ducks, and (d) geese. Values followed by different letters (age) differ significantly: capital letters α = 0.01, small letters α = 0.05 [14, 16].

Ducklings and goslings have well-developed leg muscles and poorly developed breast muscles. In Pekin ducks, the weight of muscle tissue increases 29.3-fold between 1 and 8 weeks of age. The percentage of lean meat increases from 17.1 to 30.4% of the total body weight over this period. The growth rate of breast muscles increases, and the growth rate of leg muscles decreases with age. Between 1 and 8 weeks, breast muscle weight increases 188-fold, whereas leg muscle weight only 13.6-fold. In 1-week-old ducklings, breast muscles and leg muscles account for approximately 5 and nearly 60% of total muscle weight, respectively. At 8 weeks of age, the proportion of breast muscles increases to approximately 33%, whereas the proportion of leg muscles decreases to 27% of total muscle weight (**Figure 3a** and **c**) [14].

In geese, muscle tissue weight increases 8.5-fold between 2 and 12 weeks of age. Age-related changes in lean meat weight in different carcass parts are similar in geese and ducks. The growth rate of leg muscles decreases, and the growth rate of breast muscles increases between 2 and 10 weeks of age. Over this period, the weights of breast muscles and leg muscles increase 42-fold and only 4-fold, respectively. As a result, the percentage of lean meat increases rapidly in the breast and decreases in the legs [19]. In 2-week-old geese, lean meat accounts for 18.0% of the total body weight, and it increases to nearly 29.0% at 12 weeks of age. In 2-week-old geese, leg muscles and breast muscles account for approximately 66.0% and only 7% of total muscle weight in the carcass, respectively. At 12 weeks of age, the proportion of breast muscles and leg muscles (expressed as a percentage of total lean weight) is comparable, that is, approximately 33 and 32%, respectively (**Figure 3b** and **d**) [16].

2.2.2. Skin with subcutaneous fat

Unlike Pekin ducks [9, 18] and geese [30, 31], chickens and turkeys are characterized by relatively low fat content [17, 20, 32]. The weight of fat and skin increases, at a different rate, throughout the growing period of birds. In waterfowl, subcutaneous fat can account for up to 76% of total body fat content [33]. At slaughter age, the percentage share of skin and subcutaneous fat in the total body weight is approximately 13% in broiler chickens, 10 (♂ 21 weeks) to 12% (♀ 16 weeks) in turkeys, 23% in Pekin ducks, and 19% in geese [29]. The results of research investigating different waterfowl species indicate that fat deposition as a reserve of energy is still observed in modern commercial duck and goose lines despite selective breeding aimed at eliminating atavistic features [9, 34].

In broiler chickens, the weight of skin with subcutaneous fat increases 48-fold between 1 and 10 weeks of age. Over this period, the percentage of skin and subcutaneous fat in the total body weight increases from approximately 8% in week 1 to 13% in week 3, and it remains stable until 10 weeks of age. In the analyzed period, the weight of skin and subcutaneous fat increases, but their percentage share remains at a stable level, which results from a fast growth rate of muscle tissue [13]. The distribution of skin fat and subcutaneous fat varies with age. The percentage of skin with a fat layer increases in the back (by 7%) and legs (by approx. 4%), and decreases in the breast (by approx. 4%), wings (by approx. 3%), and neck (by approx. 3.5% [17], relative to the total weight of this tissue component in the carcass. Increased fat deposition in the body cavity can also be observed in growing broiler chickens. Adipose tissue is deposited as abdominal fat, periorgan fat, and peri-intestinal fat, thus increasing the weight of slaughter offal [13].

In BIG 6 turkeys, the weight of skin and subcutaneous fat increases over 75-fold in males (from 2 to 20 weeks of age) and approximately 108-fold in females (from 2 to 20 weeks of age) [15]. Over this period, the percentage content of skin and subcutaneous fat increases by approximately 3.5% in males (from 5.4 to 8.8 %) and by 5.8% in females (from 41 to 10.9%). In males, the percentage share of skin and subcutaneous fat increases in the breast (by 7.6%), legs (by 7.9%), and back (by 1.7%), and decreases in the wings (by 11.6%) and neck (by 5.6%). In females, the percentage share of skin and subcutaneous fat increases in the legs (by 7.5%) and back (by 7.9%), decreases in the wings (by 12.1%) and neck (3.6%), and remains unchanged in the breast [20].

In Pekin ducks, between 1 and 8 weeks of age, the weight of skin with subcutaneous fat increases 25.7-fold, whereas their percentage share of the total body weight increases from 15.5% in week 1 to 24.5% in week 8. Until 3 weeks of age, the content of skin with a fat layer increases by approximately 8%, and between 3 and 8 weeks by only 1% [14]. The distribution of skin with subcutaneous fat in the carcass varies with age, with a rising tendency in the back (5% increase) and wings (4%), and a falling tendency in the neck (7% decrease) and breast (2%) [18].

In geese, the weight of skin and subcutaneous fat increases 5.4-fold from 2 to 12 weeks of age, whereas the percentage of skin with a fat layer in the total body weight remains at a similar level of approximately 19.1–19.6% [16]. The distribution of skin and subcutaneous fat in the carcass undergoes smaller changes with age than the distribution of lean meat. A rising tendency can be observed in the back (4% increase) and wings (2.5%), and a falling tendency in the breast (3.5% decrease), legs (2.5%), and neck (0.5%) [19].

2.2.3. Bones

The growth of bone tissue is completed earlier than the growth of tissue and adipose tissues. Bones are inedible components of the carcass, and the decrease in their proportion observed with age is highly desirable. However, rapid growth of muscles (in particular breast muscles) in the early life stages of birds may negatively affect their health status due to a higher incidence of leg abnormalities and deformities that result in reduced walking ability and feed utilization, followed by increased mortality [35–37].

In Ross 308 broiler chickens, between 1 and 10 weeks of age, bone weight increases approximately 23.4-fold and decreases from 11.5 to 9% when expressed as a percentage of the total body weight [13] (**Figure 2a**). Unlike the growth rates of muscle tissue and adipose tissue, the growth rate of bone tissue in different carcass parts of broiler chickens is uniform and comparable [17].

In BIG 6 turkeys, bone weight increases approximately 33-fold in males (weeks 2–20) and 34-fold in females (weeks 2–16), and the share of bones in the total body weight decreases by 4.2% in males (from 14.7 to 10.5%) and by 2.1% in females (from 13.0 to 10.9%) [15] (**Figure 2b and c**). Age-related changes are also observed in bone distribution in the carcass. In males, bone content expressed as a percentage of the total bone weight in the carcass increases significantly in the back portion (by 3.1%), and decreases in the wings (by 1.3%) and neck (by 2.0%). In females, bone content increases only in the back (by approx. 1%) and decreases in the neck (by approx. 1.5%) [20].

In Pekin ducks, between 1 and 8 weeks of age, bone weight increases 15.2-fold. Minor changes are noted in bone content—bones account for 12 and 11% of the total body weight in weeks 1 and 8, respectively [14] (**Figure 3a**). Significant age-related changes are observed in the distribution of bones in carcass parts. Until 7 weeks of age, bone content expressed as a percentage of the total bone weight in the carcass decreases in the legs (from 35.5 to 24.0%), and increases in the neck (from 9.0 to 18%) and breast (from 12% in week 2 to 16% in week 7).

The percentage of bones located in the wings, relative to the total bone weight in the carcass, remains stable (approx. 17%) [18].

In Koluda White geese, the patterns of growth in body weight and bone weight are similar until 12 weeks of age (5.3-fold and 5.2-fold increase, respectively). The proportion of bones remains at a stable level (approx. 11.9–11.5%) [16] (**Figure 4b**). Similar to Pekin ducks, also in growing geese bone content expressed as a percentage of the total bone weight in the carcass undergoes greater changes than the content of skin and subcutaneous fat. From 2 to 12 weeks of age, bone content increases rapidly in the wings (from 5.7 to 39.5%) and breast (from 9.0 to 12.9%). Until 8 weeks of age, the proportion of bones decreases rapidly in the legs (from 34.9 to 17.0%) and remains stable in the neck (approx. 9%) [19].

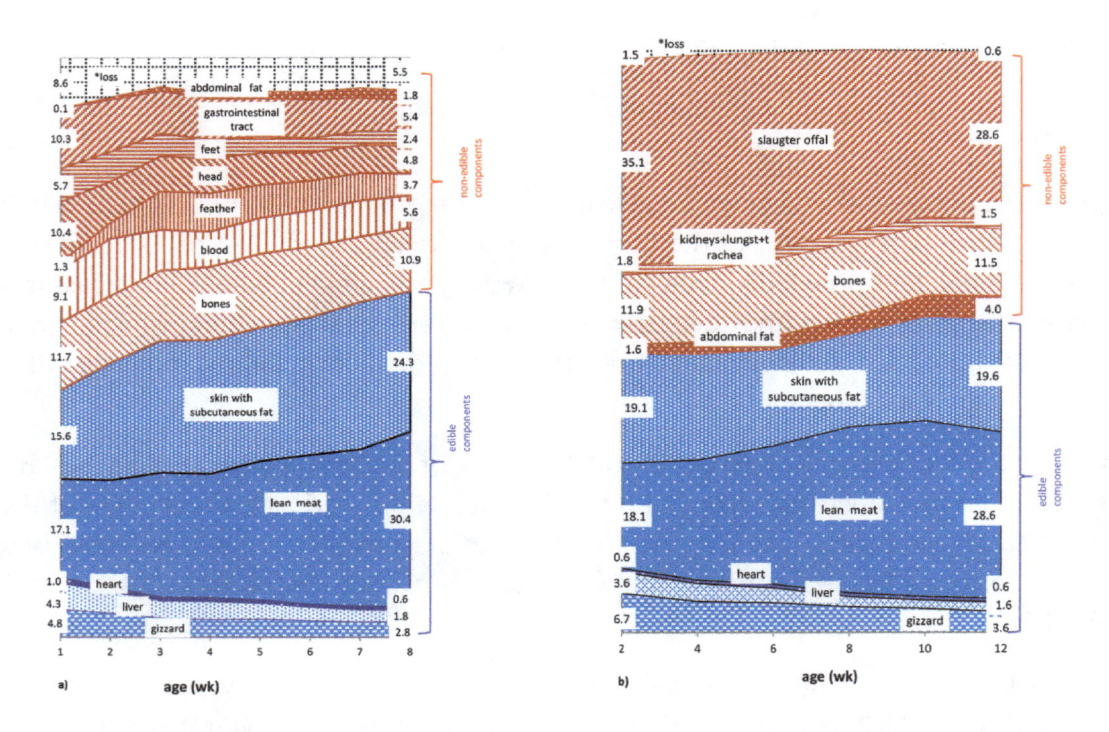

Figure 4. Percentage content of particular components in the body weight of (a) ducks and (b) geese (%). *loss = body weight loss during post-slaughter processing and dissection [14, 16].

2.3. Growth rates of organs/giblets

Full expression of the genetic potential of modern poultry lines is largely dependent on feed intake and feed conversion efficiency, which is why the composition and physical form of diets, in particular those offered to gallinaceous birds, are subject to constant modifications. Therefore, the function once performed by gastrointestinal tract segments, in particular the gizzard, has changed. In gallinaceous birds, the storage capacity and grinding activity of the gizzard have been limited. In modern broilers, the gizzard "works" less intensively due to considerable modifications in the composition and structure of feed, which is why this organ gradually diminishes in size [8]. Fast weight gain accompanied by insufficient development of some organs, for example, the heart and lungs, may lead to metabolic diseases, skeletal system diseases, and increased fat deposition [36–39].

In broiler chickens, between 1 and 6 weeks of age, heart weight increases 11.3-fold, liver weight increases 9.3-fold, and gizzard weight only 3.4-fold. In weeks 1, 6, and 10, the total weight of giblets (heart, gizzard, and liver) accounts for 8.2, 3.0, and only 2.4% of the total body weight, respectively. In 1-week-old chickens, the gizzard has a nearly 4.0% share of the total body weight, similar to the liver. Until 10 weeks of age, the percentage content of the gizzard, liver, and heart decreases over 8-fold, 2.5-fold, and 2-fold, respectively. A rapid decrease in the percentage content of the gizzard (from 4.0 to 1.5%) is noted between 1 and 2 weeks of age, and the percentage content of the liver decreases (from 3.8 to 2.1%) from 4 weeks of age. The percentage content of the heart decreases from 0.9% in week 1 to 0.4% in week 10 [13] (**Figure 2a**).

In BIG 6 turkeys, the total weight of giblets increases 13.9-fold in males (weeks 2–20) and 14-fold in females (weeks 2–16). The growth rates of individual organs vary with age. Heart weight increases 22.2-fold in males and 27.2-fold in females. Liver weight increases 15.5-fold in males and 13.5-fold in females. Gizzard weight increases 13.4-fold in males and 14.0-fold in females. At 2 weeks of age, giblets have a 5.7 and 6.1% share of the total body weight in males and females, respectively. The total percentage content of giblets decreases to 1.7% in males at 20 weeks of age, and to 2.2% in females at 16 weeks of age. The share of individual organs decreases as follows in males: gizzard, by 1.5% (from 2.2 to 0.7%); liver, by 1.0% (from 2.8 to 0.8%); heart, by 0.5% (from 0.7 to 0.3%); and in females: gizzard, by 1.3% (from 1.9 to 0.6%); liver, by 2.5% (from 3.7 to 1.2%); heart, by 0.2% (from 0.6 to 0.4%) [15] (**Figure 2b** and **c**).

According to Lilja [40], digestive organs (esophagus, gizzard, intestines, and liver) and leg muscles belong to the group of supplying organs in geese. The supplying organs are characterized by rapid embryonic development and a fast growth after hatching, which is natural since goslings need mature legs to find food and mature digestive organs to process it efficiently.

In Pekin ducks, the total weight of giblets increases 8.9-fold between 1 and 8 weeks of age. Over this period, the weight of the liver, heart, and gizzard increases 6.9-fold, 10.4-fold, and 9.5-fold, respectively. Giblet weight expressed as a percentage of the total body weight decreases from 10.1% in week 1 to 5.3% in week 8. The percentage content of individual organs decreases as follows: liver, by 2.5% (from 4.3 to 1.9%); gizzard, by 2.0% (from 4.8 to 2.8%); heart, by 0.4% (from 1 to 0.7%) [14] (**Figure 4a**).

The growth rates of visceral organs classified as giblets vary with age also in geese. Between 2 and 12 weeks of age, the total weight of giblets increases 2.8-fold, including a 2.3-fold increase in liver weight, a 2.9-fold increase in gizzard weight, and a 4.7-fold increase in heart weight. A statistically significant increase in liver weight, gizzard weight, and heart weight is observed until 4, 6, and 8 weeks of age, respectively. The percentage content of giblets in the total body weight decreases with age, from 10.9% in week 2 to 5.8% in week 12. Over this period, the share of the gizzard and liver decreases by 3.1% (from 6.7 to 3.6%) and 2.0% (from 3.6 to 1.6%), respectively, whereas the share of the heart in the total body weight remains stable at around 0.6% [16] (**Figure 4b**).

2.4. Growth rate of the slaughter offal portion

During post-slaughter processing, offal is separated from the carcass. In poultry, the slaughter offal portion includes blood, feathers, feet, head, trachea, lungs, intestines (including the contents and peri-intestinal fat), spleen, pancreas, testes, components of the female reproductive system, and abdominal fat. In some regions of the world, abdominal fat and feet are classified as edible components, but they are generally considered as offal. Contemporary consumers prefer low-fat products; therefore, abdominal fat is treated as slaughterhouse waste [41]. However, the fat of waterfowl is used in regional cuisines for the production of specialty foods. Geese have higher abdominal fat content than other poultry species [31]. Fat deposition in the lower parts of the body in waterfowl is a natural adaptation to the environmental conditions. Duck and goose fat is considered to be healthier than pork fat, due to higher concentrations of essential unsaturated fatty acids [33, 42]. However, some researchers emphasize the lower durability of the former. In ducks and geese, the abdominal fat weight increases over the entire growing period, thus increasing both the weight of undesirable offal and overall production costs. Feathers are a by-product of poultry production. Today, the economic importance of the feathers of gallinaceous birds is relatively low [3]. Plumage development is particularly important in ducks and geese because carcass quality deteriorates during the feathering process, which affects the processing suitability of raw material. Feathers account for approximately 4.2, 4.0–4.5, and 5.4% of the total body weight in turkeys, chickens and ducks, and geese at slaughter age, respectively. The rate of feathering is species-specific, affected by nutrition and determined genetically [43].

In broiler chickens, the total weight of slaughter offal increases 17-5-fold between 1 and 10 weeks of age, including a 83-fold increase in feathers weight, a 41-fold increase in abdominal fat weight, a 19-fold increase in the weight of feet, a 14-fold increase in blood weight, an 11-fold increase in the weight of the gastrointestinal tract, and a 10-fold increase in head weight. The growth rates of the above components are slower than the whole-body growth rate, and their percentage content decreases with age, from 33.0% to approximately 19% in broiler chickens. The proportions of most offal components decrease during the growing period, except for abdominal fat and feathers (3.2 and 1.8% increase, respectively). Between 1 and 10 weeks of age, the smallest changes are noted in the percentage content of feet (2% decrease, from 4.5 to 2.5%), and the greatest changes are observed in the percentage content of the gastrointestinal tract (8.5% decrease, from 12.1 to 3.6%). The carcasses of 1-week-old chickens contain low amounts of abdominal fat, whereas at 10 weeks of age abdominal fat content reaches 3.2% and is almost equal to that of gastrointestinal tract [13] (**Figure 2a**).

In male turkeys, the weight of slaughter offal increases 23.5-fold between 2 and 20 weeks of age. Over this period, blood weight increases approximately 35-fold, feathers weight increases approximately 31-fold, and the combined weight of the lungs and trachea nearly 38-fold. The growth rates of the remaining nonedible components are slower; gastrointestinal tract weight increases only 14.5-fold. The percentage content of slaughter offal decreases in males from

30.1% at 2 weeks of age to 15.8% at 20 weeks of age. The greatest decrease is noted in the percentage share of the gastrointestinal tract (by approx. 7%), head (3%), and feet (2%). In female turkeys, the weight of slaughter offal increases 20-fold between 2 and 16 weeks of age. The fastest growth rate is observed in feathers weight (approx. 64-fold increase), followed by the weight of feet (29-fold), lungs and trachea (27-fold), and blood (25-fold). Similar to males, the growth rate of the gastrointestinal tract is relatively slow (approx. 10-fold increase). The percentage content of slaughter offal decreases in females from 38.5% at 2 weeks of age to 18.0% at 16 weeks of age. Age-related changes in the content of individual offal components are similar to those noted in males [15] (**Figure 2b** and **c**).

In Pekin ducks, offal content expresses as a percentage of the total body weight decreases from 37% in week 1 to 24% in week 8. The weights of feathers and abdominal fat change to the greatest extent. Between 1 and 8 weeks of age, feathers weight increases approximately 65-fold. The carcasses of 1-week-old ducks contain low amounts of abdominal fat, but its weight increases nearly 200-fold by 8 weeks of age. The growth rates of the remaining nonedible components are slower; from 1 to 8 weeks of age, the weights of blood, head, feet, and gastrointestinal tract increase 11-fold, 7.7-fold, 7.2-fold, and 9.1-fold, respectively. Changes in the growth rates of the above components and the body weight of birds contribute to an increase in the percentage content of feathers and abdominal fat (by 2.4 and 1.5%, respectively), and a decrease in the percentage content of gastrointestinal fat (5%), blood (3.3%), head (5.4%), and feet (3.3%) [14] (**Figure 4a**).

In geese, between 2 and 12 weeks of age, the total weight of slaughter offal increases 4.2-fold or 4.6-fold, depending on the classification of abdominal fat as an edible or inedible component, respectively. Over this period, abdominal fat weight increases approximately 13.6-fold. The increase in offal weight in geese aged 2–12 weeks is not uniform. The total weight of slaughter offal increases rapidly (2.9-fold) in weeks 2–4, it stabilizes and remains relatively unchanged until week 10, and increases again in weeks 10–12. The main reason for the increase in slaughter offal weight in geese aged 10–12 weeks is enhanced peri-intestinal fat deposition. Despite the increase in slaughter offal weight, its proportion in the total body weight decreases from 35% in week 2 to 29% in week 12 [16] (**Figure 4b**).

3. The effect of age on the percentage content of edible and nonedible components

The total weights of edible and nonedible components in the carcasses of different poultry species are an important economic consideration since edible carcass parts are used as raw materials for the manufacture of processed meat products, and the waste load from meat-processing plants has to be effectively managed or disposed of. In view of the fact that global poultry meat production continues to increase, the ratio between edible and inedible weights in poultry carcasses deserves particular attention [2, 3, 7]. The whole-body growth rate of birds varies over time, similarly as the growth rates of internal organs and tissue components, which affects the proportions between edible and nonedible components in the carcass. Research results show that the rate of age-related changes in the content of edible and nonedible carcass portions varies across poultry species [13–16].

3.1. Percentage content of the edible portion

In chickens slaughtered at 1 week of age, the ratio between edible and nonedible parts is estimated at 1:1, whereas a more desirable value of 2.2:1 is noted in birds slaughtered at 10 weeks of age. From weeks 1–10, the total weights of edible and nonedible components in broiler carcasses increase 42-fold and only 20-fold, respectively. In 1-week-old broiler chickens, edible components account for approximately 47.0% of the total body weight, and their content increases to 63.0 and 67% at 6 and 10 weeks of age, respectively [13] (**Figure 2a**).

Similar to chickens, age-related changes in the weights and percentages of edible and nonedible components are also observed in turkeys. In males, between 2 and 20 weeks of age, the weights of edible and nonedible portions increase 63.5-fold and approximately 26.5-fold, respectively. In 2-week-old males, edible components account for nearly 52% of the total body weight, compared with 72% in 20-week-old birds. In the group of edible components, muscle tissue is characterized by the highest growth rate (40.4% in week 2, 60.9% in week 20), followed by skin and subcutaneous fat (5.4% in week 2, 8.8% in week 20). The content of giblets, expressed as a percentage of the total body weight, decreases with age (in contrast to the remaining edible components), from 5.7% in week 2 to 1.8% in week 20. A similar trend can be observed in female turkeys. Until 16 weeks of age, edible and inedible weights increase approximately 63.7-fold and 23.7-fold, respectively. The share of muscle tissue in the total body weight increases from 33.7% in week 2 to 55.6% in week 16, the share of skin with subcutaneous fat increases from 4.1% to 10.9%, and the share of giblets decreases from 6 to 2.4% [15] (**Figure 2b** and **c**).

In 1-week-old ducks, the weight of edible components is lower than the weight of nonedible components (0.8:1). In Pekin ducks, between 1 and 8 weeks of age, edible and inedible weights increase 23-fold and 12.1-fold, respectively, and the ratio between edible and nonedible components reaches 1.6:1 in 8-week-old birds. Edible components account for 42.8% of the total body weight in week 1 and for 59.9% in week 8. The most significant changes are observed in the percentage content of muscle tissue (10.5% increase) and giblets (5.1% decrease). The share of skin and subcutaneous fat remains stable at 19.3% throughout the growing period. As a result, the proportions between tissue components change. In 1-week-old ducks, muscle tissue, skin with a fat layer, and giblets account for 40, 37, and 24% of total edible weight, respectively. More desirable values are noted in 8-week-old birds, where muscle tissue, skin with fat, and giblets account for 50, 41, and only 9% of total edible weight, respectively [14] (**Figure 4a**).

In geese, the ratio between edible and inedible weights is 1:1 at 2 weeks of age, and 1.4:1 (variant I, abdominal fat classified as an edible component) or 1.2:1 (variant II, abdominal fat classified as a nonedible component) at 12 weeks of age (Murawska 2013b). The most valuable edible component, muscle tissue, has an 18% share of the total body weight in week 2 and 29% in week 12. Between 2 and 12 weeks of age, the percentage content of skin and subcutaneous fat in the total body weight of geese remains stable at 19.5%, whereas the share of giblets decreases from 10% in week 2 to 6% in week 12 [16] (**Figure 4b**).

3.2. Percentage content of the nonedible portion

The nonedible portion of poultry carcasses comprises slaughter offal and bones. The content of those components, expressed as a percentage of the total body weight of birds, decreases with age.

In 1-week-old broiler chickens, nonedible components account for 45.3% of the total body weight on average, and their content decreases to 34.4 and 30.2% at 6 and 10 weeks of age, respectively. Over this period, a greater decrease is observed in the share of slaughter offal (by 12.5%, from 33.7% in week 1 to 21.2% in week 10), in comparison with bones (by 2.5 %, from 11.6 in week 1 to 9.1% in week 10) [13] (**Figure 2a**).

In 2-week-old turkeys, nonedible components have a 44.7 (♂) and a 49.9% (♀) share of the total body weight, and their content decreases to 26.0 (♂, week 20) and 29% (♀, week 16) at slaughter age. Similar to broiler chickens, slaughter offal content decreases by over 15% (from 29.7 to 14.5%) in males and by 19% (from 36.9 to 17.9%) in females, whereas bone content decreases at a slower rate (♂ 4.2% decrease, from 14.7 to 10.5%; ♀ 2.1% decrease, from 13.0 to 10.9%) [15] (**Figure 2b** and **c**].

In ducks, nonedible components account for 48.6% of the total body weight in week 1, and decrease to 34.6% in week 8. Slaughter offal accounts for 39.7 and 27.2% of the total body weight in weeks 1 and 8, respectively. Bone content decreases from 11.7% in week 1 to 10.9% in week 8 [14] (**Figure 4b**).

In geese, between 2 and 12 weeks of age, the percentage content of inedible components decreases mostly due to a decrease in the share of slaughter offal (approx. 6.5%) because the proportion of bones remains at a stable level (approx. 11.9–11.5%). In variant I (abdominal fat classified as an edible component), the share of nonedible parts decreases from 48.08 to 41.6%, and in variant II (abdominal fat classified as a nonedible component) from 50.2 to 45.4% [16] (**Figure 4b**).

4. Conclusion

The research findings presented in this chapter can be used in an analysis of potential health threats resulting from the undesirable differences between the growth rates of individual organs and the whole-body growth rate of birds. The relationship between the growth rates of muscle tissue and adipose tissue is also an important consideration for carcass quality and poultry production efficiency. Due to the specific structure of subcutaneous adipose tissue in poultry, analyses of carcass fat content should include both skin with a subcutaneous fat layer and abdominal fat. Information on age-related changes in the weight and content of edible and inedible carcass components may contribute to the effective processing of poultry meat and the effective management of slaughter offal, including bones separated from muscle tissue in the mechanical deboning process.

A prolonged rearing period contributes to higher final body weight and higher muscle yield, but overall production costs increase because older birds are characterized by a slower growth rate and a higher feed intake per kg of body weight gain. Those factors should be taken into account when determining the optimal slaughter age.

Author details

Daria Murawska

Address all correspondence to: daria.murawska@uwm.edu.pl

Department of Commodity Science and Animal Improvement, Faculty of Animal Bioengineering, University of Warmia and Mazury in Olsztyn, Olsztyn, Poland

References

[1] Shahin KA. Sources of shared variability of the carcass and non-carcass components in Pekin ducklings. Annales de Zootechnie. 2000;49:67–72.

[2] Meeker DL, Hamilton CR. An overview of the rendering industry. In: Meeker, editor. 16 Essential rendering. National Renderers Association, Virginia; 2006. ISBN: 0-9654660-3-5. p. 1–17. ch1

[3] Kornillowicz-Kowalska T, Bohacz J. Biodegradation of keratin waste: Theory and practical aspects. Waste Management. 2011;31:1689–1701.

[4] Wood BJ. Calculating economic values for turkeys using a deterministic production model. Canadian Journal Animal Science. 2009;89:201–213.

[5] Case LA, Miller SP, Wood BJ. Determination of the optimum slaughter weight to maximize gross profit in a turkey production system. Canadian Journal Animal Science. 2010;90: 349–356.

[6] Case LA, Miller SP, Wood BJ. Factors affecting breast meat yield in turkeys. World's Poultry Science Journal. 2010;66:189–201.

[7] Plumber HS, Kiepper BH. Impact of poultry processing by-products on wastewater generation, treatment and discharges. In: Proceedings of the Georgia Water Resources Conference (GWRC'11); USA, GA; 11–13 April, 2011. Available from: http://www.gwri.gatech.edu/sites/default/files/files/docs/2011/6.5.3Plumber.pdf

[8] Havenstein GB, Ferket PR, Grimes JL, Qureshi MA, Nestor KE. Comparison of the performance of 1966-versus 2003-type turkeys when fed representative 1966 and 2003

turkey diets: growth rate, livability, and feed conversion. Poultry Science. 2007;86:232–240.

[9] Fan HP, Xie M, Wang WW, Hou SS, Huang W. Effects of dietary energy on growth performance and carcass quality of white growing Pekin ducks from two to six weeks of age. Poultry Science. 2008;87:1162–1164.

[10] Yegani M, Korver DR. Factors affecting intestinal health in poultry. Poultry Science. 2008;87:2052–2063.

[11] Murawska D, Bochno R, Michalik D, Janiszewska M. Age-related changes in the carcass tissue composition and distribution of meat and fat with skin in carcasses of laying-type cockerels. Archiv für Geflügelkunde. 2005;69:135–139.

[12] Ristic M, Damme K, Freudenreich P. Carcass value of ducks and geese in dependence of breed and age. Fleischwirtschaft. 2006;86:107–110.

[13] Murawska D, Kleczek K, Wawro K, Michalik D. Age-related changes in the percentage content of edible and non-edible components in broiler chickens. Asian-Australasian Journal Animal Science. 2011;24:532–539.

[14] Murawska D. The effect of age on the growth rate of tissues and organs and the percentage content of edible and non-edible carcass components in Pekin ducks. Poultry Science. 2012;91:2030–2038.

[15] Murawska D. Age-related changes in the percentage content of edible and non-edible components in turkeys. Poultry Science. 2013;92:255–264.

[16] Murawska D. The effect of age on the growth rate of tissues and organs and the percentage content of edible and inedible components in Polish Koluda White Geese. Poultry Science. 2013;92 :1400–1407.

[17] Bochno R, Brzozowski W, Murawska D. Age-related changes in the distribution of meat, fat with skin and bones in broiler chicken carcasses. Polish Journal of Natural Science. 2003;14:335–345.

[18] Bochno R, Brzozowski W, Murawska D. Age- related changes in the distribution of lean, fat with skin and bones in duck carcasses. British Poultry Science. 2005;46:199–203.

[19] Bochno R, Murawska D, Brzostowska U. Age-related changes in the distribution of lean, fat with skin and bones in goose carcasses. Poultry Science. 2006;85:1987–1991.

[20] Murawska D, Kozłowski K, Tomaszewska K, Brzozowski W, Zawacka M, Michalik D. Age-related changes in the tissue composition of carcass parts and in the distribution of lean meat, fat with skin and bones in turkey carcasses. European Poultry Science. 2015;ISSN:1612–9199. DOI: 10.1399/eps.2015.103.

[21] Murawska D, Bochno R. Age-related changes in the percentage content of tissue components in geese. Journal Central European Agriculture. 2008;9:211–216.

[22] Laughlin K. The evolution of genetics, breeding and production. Temperton Fellowship. University College, Newport, Shropshire. 2007; Report No. 15:1–55.

[23] Groeneveld LF, Lenstra JA, Eding H, Toro MA, Scherf B, Pilling D, Negrini R, Finlay EK, Jianlin H, Groeneveld E, Weigend S, GLOBALDIV Consortium. Genetic diversity in farm animals-a review. Animal Genetics. 2010;41(Suppl. 1):6–31. doi: 10.1111/j. 1365-2052.2010.02038.x.

[24] Lesage L, Gauthier G. Growth and organ development in greater snow goose goslings. The Auk. 1997;114:229–241.

[25] Książkiewicz J. Mint: a mallard duck- *Anas platyrhynchos* L., means a flat-billed duck. Wiadomości Zootechniczne. 2006;1:25–30.

[26] Gordy JF. Broilers. In Skinner JL, editor. American Poultry History, 1823–1973. Madison, WI: American Poultry Historical Society, 1974, p. 370–432.

[27] Aviagen 2016, Parent Stock management manual Ross308. Available from: http:/www.aviagen.com [Accessed: 2016-04-28].

[28] Velleman SG, Anderson JW, Coy CS, Nestor KE. Effect of selection for growth rate on muscle damage during turkey breast muscle development. Poultry Science. 2003; 82:1069–1074.

[29] Jankowski J, Kozłowski K. Breeding and use of turkeys. In: Jakowski J, editor. Breeding and use of poultry. 1st ed. Warsaw: PWR i L; 2012. ISBN: 978-83-09-01142-2. p. 305–354.

[30] Gumulka M, Wojtysiak D, Kapkowska E, Połtowicz K, Rabsztyn A. Microstructure and technological meat quality of geese from conservation flock and commercial hybrids. Annals of Animal Science. 2009;9:205–213.

[31] Liu BY, Wang ZY, Yang HM, Wan JM, Xu D, Zhang R, et al. Influence of rearing system on growth performance, carcass traits and meat quality of Yangzhou geese. Poultry Science. 2011;90:653–659.

[32] Rivera-Torres V, Ferket PR, Sauvant D. Mechanistic modeling of turkey growth response to genotype and nutrition. Journal Animal Science. 2011;89:3170–3188.

[33] Pingel H. Improvement of carcass and meat quality of ducks and geese, Fleischwirtschaft. 2006;86:101–105.

[34] Tsuji LJS, Martin ID, Martin ES, LeBlanc A, Dumas P. Spring-harvested game birds in the western James Bay region of northern Ontario, Canada: the amount of organochlorines in matched samples of breast muscle, skin, and abdominal fat. Environmental Monitoring and Assessment. 2008;146:91–104.

[35] Bessei W. Welfare of broilers: a review. World's Poultry Science Journal. 2006;62:455–466.

[36] Charuta A, Cooper RG, Pierzchała M, Horbańczuk JO. Computed tomographic analysis of tibiotarsal bone mineral density and content in turkeys as influenced by age and sex. Czech Journal Animal Science. 2012;57: 572–578.

[37] Charuta A, Dzierzecka M, Komosa M, Kalinowski L, Pierzchała M. Age- and sex-related differences of morphometric, densitometric and geometric parameters of tibiotarsal bone in Ross Broiler chickens. Folia Biologica. 2013;3–4:211–220.

[38] McKay JC, Barton NF, Koerhuis ANM, McAdam J. Broiler production around the world. XXI. In: Proceedings of the World's Poultry Congress; 20–24 August 2000; Montreal. Canada: 2000. p. 20–24.

[39] Gesek M, Szarek J, Otrocka-Domagała I, Babińska I, Paździor K, Szweda M. et al. Morphological pattern of the livers of different lines of broiler chickens during rearing. Veterinarni Medicina. 2013; 58:16–24.

[40] Lilja C. Postnatal growth and organ development in the goose (*Anser anser*). Growth. 1981;45:329–341.

[41] Kostecka M. Characterization of poultry fat and rapeseed oil mixture prior to and after the enzymatic interesterification. Żywność. Nauka.Technologia. Jakość. 2008;5:257–272.

[42] Chartrin PM, Bernadet D, Guy G, Mourot J, Duclos MJ, Baéza E. The effects of genotype and overfeeding on fat level and composition of adipose and muscle tissues in ducks. Animal Research. 2006;55:231–244.

[43] Leeson S, Walsh T. Feathering in commercial poultry II. Factors influencing feather growth and feather loss. World's Poultry Science Journal. 2004;60:52–63.

Poultry Litter Selection, Management and Utilization in the Tropics

Musa I. Waziri and Bilkisu Y. Kaltungo

Abstract

In many places, poultry farms are sited and intensively managed in the residential areas with little welfare improvement and major concern of poultry waste disposal. The mean poultry litter generated per bird/day was estimated at 0.11 kg so that millions of metric tons of poultry manure are annually generated. Wastes generated from poultry farms constitute hazard to public health as well as potential source of conflict between neighbours. Beside environmental consequences, poultry waste impacts serious welfare and production effects on poultry. Suitable poultry droppings and moist absorbents referred to as litter materials were later discovered. Recently, conventional caging of birds is considered unethical, common litter materials are seasonally available, wood based litter materials are now being diverted for the manufacture of other wood products, the use of poultry litter as fertilizer and livestock feed supplement has increased, therefore, the demand and price for litter materials is now magnified. Adequate litter materials cannot be easily met by farmers and non-environmentally friendly alternative litter materials may be sought by farmers creating negative socio-economic impacts on poultry and the environment. Therefore, careful selection, adequate management and proper storage and utilization of poultry litter are here given due attention.

Keywords: Poultry, Litter, Selection, Management, Utilization, Tropics

1. Introduction

Individuals and organizations worldwide rely directly or indirectly on the poultry industry for substantial portion of their income and low-cholesterol animal protein intake [1, 2]. Breeders and broilers in most countries are raised exclusively on deep litter system, whereas layers in many places are initially raised on deep litter before transfer to cage [3, 4]. The poultry industry has over the years rapidly expanded with increasing concern of poultry waste disposal [2, 5]. Ammonia and greenhouse gases produced by poultry litter impact negatively on the environment [6]. Therefore, environmentally friendly and economically sustainable technologies for poultry waste management must be focused in today's climatic challenges. Efforts to manage poultry wastes under intensive production systems led to the discovery of suitable poultry droppings and moist absorbents called litter material [7, 8]. Animal welfare organizations have seriously raised concern on how commercial birds are deprived of exhibiting their natural behaviours by caging. To this end, some advanced countries have enacted laws banning the use of conventional poultry cages. Wood shavings and saw dusts are now being utilized to produce paper and other wood products while alternative litter materials, such as rice hulls, pines and groundnut shells, are seasonal [8, 9]. These imply that, deep litter system of poultry management may be more acceptable than caging and consequently the demand and cost of litter materials will significantly increase [9]. Adequate litter materials in deep litter management system may therefore not be guaranteed and/or the litter materials will become scarce and not easily accessible [4]. The implication is that poultry farmers are not likely to obtain adequate and good quality litter material for their birds [1, 8]. This may result in farmers trying many unconventional litter materials that may affect poultry health, welfare, performance and the environment.

Some of the economic losses associated with poor litter in poultry include musculoskeletal problems of the foot and leg, increased respiratory tract infections and poor production performance as a detriment to low feed consumption utilization [9]. Dan et al. observed that the cost of poultry waste disposal is normally omitted in casting production expenses which often contribute significantly to the overall production expenses. In some countries where litter is properly managed, old litter is removed and replaced with new one after many sets of birds are cropped [10, 11]. This seems not to be possible in many developing countries where poultry houses are poorly constructed coupled with poor hygienic and management practices and frequent weather variations which may require frequent litter change and/or adjustment of poultry pen environment. Many poultry farmers in such nations do not make use of facilities required to determine ammonia, humidity, temperature and ventilation levels and efficiencies in poultry pens. Ventilation has been reported to be the primary way to reduce or eliminate moisture in poultry houses, and temperature determines the degree of litter caking during cool weathers [12].

Poultry litter contains high nitrogen and phosphorus making it a very good organic fertilizer and feed supplement [13–17]. The poultry manure enhance physical, chemical and biological fertility of soil by ensuring adequate levels of organic matter, water holding ability and oxygen diffusion rates. However, poultry litter may contain pathogenic microorganisms, drug residues and hard or metallic objects that are injurious to crops, poultry, humans and other domestic animals [8, 11, 17–19]. Arsenic compounds are known for their potentials to cause cancers but unfortunately used to control coccidiosis in many countries [20]. This study discusses aspects of selection, proper management, storage and efficient utilization of poultry litter.

2. Selection of poultry litter materials

Many research works were conducted to determine the suitability of wood shavings, sand, pine peanut shells, shavings, shredded papers or paper chips, dry straw, rice hulls, maize cobs, corn silage and peat as alternative litter materials [7, 21, 22]. Timber by-products, rice hulls and shredded papers appear to be most accessible worldwide. The basic requirements of a good litter include moisture holding capacity, microbial tolerating ability, low cost, availability and non-toxicity to poultry [23, 24]. Bases for choosing good litter materials should include the ability to protect birds from dirt, damp and cold floor; it should also be able to adequately conserve heat and absorb moisture. For good production performance, litter materials should provide comfort for birds [24, 25].

3. Wood shavings

Soft or hard wood shavings from different trees are available year round in many places. These are normally obtained from the wood work and furniture enterprises. Wood shavings sourced from soft wood the best litter material but is highly demanded for making paper, fibreboard and cardboard making it difficult to obtain [10]. In some instances, the woods may be treated with some chemical preservatives like copper chrome arsenate or even organophosphates which may be harmful. Wood shaving is the most common poultry litter material which today is characterized by periodic shortage due to increasing number of poultry producers [8, 26]. Unfortunately, hard wood shavings are reported to poorly absorb moisture and are frequently contaminated with *Aspergillus* [9], showed highest prevalence of Salmonella organism [22] and posed a significant problem when obtained from chemically treated woods [4]. Wood shaving will be an ideal litter material if free of contaminants and if properly managed. Some countries are reported to specifically produce uniformly sized untreated soft wood shaving as litter material for poultry. Below is a plate (**Plate I**) showing a combination of soft and hard wood shavings used as litter material in many parts on Nigeria.

Plate I. A mixture of soft and hard wood shavings used as poultry litter in Zaria, Nigeria.

Saw dusts are finer wood materials from wood work and furniture enterprises and used as alternative litter material especially where wood shaving is unavailable or unaffordable. Unfortunately, poultry especially chicks consumes as much as 4% of their diet as litter. Turkeys are more prone to consume more litter than chickens, which may lead to nutritional deficiency, crop impaction, starvation and subsequent mortality [9, 24]. Sawdust is very popular as litter material in many places, but it regularly cakes especially around drinkers and feeders [9, 27]. Sawdust has high moisture holding ability but is commonly contaminated with *Aspergillus* [9].

Plate II. Saw dust used as poultry litter material in Zaria, Nigeria.

The above plate (**Plate II**) shows saw dust utilized in many parts of Nigeria as poultry litter. It is common and less expensive than saw dust.

Rice hulls have been reported to work well as litter material because of their uniform sizes, less dusty and most importantly have high thermal conductivity, drying rate and are compressible [4]. They serve as very good litter material for broiler chickens and its organic-based manure has been reported to be well suited for gardeners. Rice hulls can be better litter when combined with other litter materials like pine and wood shavings. In some tropical regions where rice is annually cultivated, rice hulls may be costly and only available seasonally and could also be restricted to certain regions where rice is grown. Unfortunately, rice hulls can easily mould and bacterial growth has been found to be common thus restricting its use [9, 24].

3.1. Corn silage litter

This as litter material that was found to be a suitable alternative for other common litter materials with an added advantage of very low Salmonella prevalence [22]. However, it is not popular and not always available because corn is mainly a seasonal crop grown in certain restricted areas of the tropics.

3.2. Straw litter

This is any grass material that can directly be used or processed to be used as (straw litter) material. It is used worldwide especially in cereal producing countries. Wheat straw is the most popular but rye straw was proven to be most superior over other grass straws. Straw generally dries quickly thus discouraging fungal growth, however, it is difficult to manage because of its length and can easily cake and balls up when squeezed requiring routine litter readjustment [20, 24]. However, for straw to be an efficient litter material, it should be chopped to one inch or 37 mm pieces to prevent long straw its usual ability to bridge or mat quickly [24]. Despite the fact that chopped straw was found to be free of Salmonella, it had side effects on environmental health [22]. Straw litter can serve better and more effectively when used as a top dressing over old litter. It is cheaper and most readily available litter material and therefore more economically viable [24]. Straw can best be used in 1:1 combination ratio with shavings, rice hulls or old litters [24].

3.3. Bagasse

Is a sugarcane by-product when sugar or the juice from sugarcane is extracted. It is common but not readily used as litter in many parts of northern Nigeria where sugar cane is cultivated. It is normally burnt and taken to farms as organic fertilizer. If properly harnessed, it can be a good litter material because it is highly moisture absorbent and dries easily; however, care must be taken as it cakes easily [24].

3.4. Recycled papers

Many paper products, such as newspapers, cardboard, shredded papers and chopped newspapers, are increasingly being used as alternative litter materials in poultry farms because they are relative cheap and available. Unfortunately, paper products are reported to retain high level of moisture. This increases its ability to cake easily enabling breast blister formation and other carcass defects [4, 8, 22, 24, 25]. When recycled papers are to be used as litter material, they are best applied as top litter dressing or they can be mixed with other conventional wood-based litter materials [8, 24]. The best size for a shredded paper as litter material is 1–2 cm in diameter, and old newsprint papers are often recommended to be used because some printing inks are toxic to chicks, whereas glossy papers are reported not to absorb moisture [12].

Pine shavings as litter materials were successfully experimented in which variations on broiler performance were not observed [7].

Sand as litter material did not show any variation in broiler performance when compared to other conventional litter materials. Sand was shown to reduce darkling bee infestations, and it had longer period of up to 5 years before clean-out. It was shown not to heat up appropriately during cold periods when compared to wood shavings and is therefore best suited as litter material during the summer [7, 24]. Sand was also shown to be less dusty and improved foot pad quality. However, sand was found not to be compatible with composting, incinerating and pelleting but is currently attracting research interest in many places [7, 24]. Because of its availability, less cost and accessibility it may stand to be the most acceptable to the common man.

Composted litter is another cheap, dust-free litter material that is often associated with low odour and low pathogenic organisms or parasites and therefore appears to be good and suitable litter, however, it is not recommended as litter material during brooding because NH_3 in composted sand litter may persist for long [4]. Composted sand litter contains higher manure proportion for crop application and also decomposes faster when applied to crops. Furthermore, the pH of the manure from composted sand litter ranges between 5.5 and 6.5, which is most suitable for vegetables and fruits [12].

Groundnut or peanut shells as shown in **Plate III** below have been used alone or in combination with other litter materials where groundnut is cultivated in Nigeria. It only seems to protect birds from being in direct contact with the floor as its warming and moisture absorbing abilities are poor. It may, however, be obtained at no cost but its suitability as a good litter material needs further evaluation.

Plate III. Peanut/groundnut shells used as poultry litter material in some parts of Nigeria.

Finally, each type of poultry bedding material is subject to factors that will enable it to be a successful litter material [9].

4. Litter management

Organic manure decomposition produces odorous gases, such as amides, amines, mercaptans, sulphides and disulphides. These biogases may irritate and disrupt the respiratory tract epithelial lining of animals and men leading to high degree of susceptibility to respiratory tract infections [8].

4.1. Basic factors of consideration in litter management

4.1.1. Ventilation

Is a major factor in moisture control in poultry houses because it provides adequate air movement that will enhance moisture evaporation in poultry houses. Decreased litter moisture will subsequently lead to decrease in free NH_3 and CO_2, which may lead to increased air dust content in poultry houses [8].

4.1.2. Temperature

It plays a significant role in influencing clumping into layers of litter materials referred to as caking. A normal litter is expected to break easily if hand squeezed but will remain moulded if the moisture content is high and if temperature is high it will result into caking.

4.1.3. Water spillage

It is difficult or almost impossible to avoid water spillage when using manual drinkers in poultry managed on deep litter system. In semi-automated systems of management, water lines that make use of nipple drinking points will have some chances of water spillage and the water lines with nipple and cups will have minimal water spillage.

4.1.4. Quantity of the litter material used

If litter material is liberally applied to a required thickness, the litter will sufficiently absorb moisture making birds more comfortable and enabling them to exhibit some of their natural behaviours thus enhancing birds' growth, performance, environment and welfare [12].

Good litter is made of adequate materials applied to a sufficient depth of at least 2 cm for cool sand, 5–10 cm in case of wood shavings, 10 cm for chopped straw and generally 8 cm for any other litter material on a damp-free floor [2, 7, 8]. However, environmental, management and indoor conditions of poultry houses especially housing temperatures, stocking density and air movements have significant influences on litter quality and degree of NH_3 emissions [6, 8, 28]. Concentrated wastes in the form of uric acid passed out by birds make it possible to house many birds on litter with a major challenge of moisture control [7, 8, 24]. Therefore, deep litter management to miniimze dampness is necessary but seems not to be given due attention in the Nigerian poultry industry [8, 29]. Efforts to maintain good litter should therefore consider factors, such as type of material used as the litter depth, the season of the year, the depth requirement, stocking density, watering devices, nature of the floor, provided pen ventilation, routine litter management practices, litter amendment facilities and procedure at disposal and incidences of litter-related diseases [7]. Generally, the depth and type of the litter used varies with the type of litter material available. This should enhance but not retard performance [2, 7]. High stocking density leads to humid environment due to decreased water and gas exchange between air and litter [6] there is usually high chance of feed and water spillage due to space competition, high levels of waste secretions and excretions into the litter that will lead to temperature and ammonia build up in the poultry house and subsequently high chances of bad litter occurrence. Well-ventilated poultry houses with relatively light stocking densities will maintain good litter [8, 30]. Litters if well managed can be changed at the end of each cycle [8, 30], unless the litter appears bad or if diseases outbreaks occur. Dusty, wet and cakey litters are signs of badly managed litter [8, 30]. A good litter should adhere slightly when squeezed, it should easily break when dropped from the hand, but when litter is too wet it balls up if squeezed in the hand, too dry litter does not normally adhere [30]. Litter materials on earthen floors hold as much as 10% moisture making it almost impossible to effectively manage than litters on damp-proofed concrete floors [28]. A quick test for litter dampness will be if the back of the hand feels damp when applied onto a litter, then it possibly contains at least 30% moisture which rapidly converts uric acid to toxic ammonia, supports the growth of fly larvae and coccidian organisms and also encourages breast blisters [31].

The rules of litter management are few but most decisions are subject to operator's judgment. Some of these emphasize that litter materials should be checked for bacterial and fungal contaminations, litter materials of fine particle should be covered with paper to avoid litter consumption not eating and new litter materials should be treated with approved anti-fungal agents while litter intended to be reused should also be treated with lime [8, 31]. In managing litter, special attention should be paid to drinker points because such areas are normally damp liable to caking as shown in **Plate IV**, such points should be turned and tilled to activate litter or be removed and fresh litter material added [8, 31, 32]. Unfortunately, tilling of litters is frequently associated with rapid increase in ammonia levels in poultry houses; this should be done with windows open or fans on to rapidly dissipate the ammonia [31]. A good working litter gives desirable warmth and cold to the poultry house while a wet litter cools the house due to heat loss in the process of drying out [25]. It is dangerous to the birds and the operator to allow ammonia build beyond 40 part per million (PPM) is poultry houses [33]. This consequently will lead to decreased feed intake and productivity, respiratory tract infections and blindness [34]. However, ammonia levels of 15–20 PPM is acceptable and can be estimated fairly accurately using the operator's sense of smell or litmus paper or more accurately using commercially available dragger gas detector [35].

Plate IV. Inadequate and poorly managed litter around a drinker in Zaria, Nigeria.

5. Management of ammonia in poultry farms

Some micro-organisms present in the litter convert birds' excreta and spilled feeds to ammonium (NH_4^+), which is soluble in water and is convertible to ammonia in the presence of high pH and temperature [32]. However, a high ammonia level in litter is desirable in increasing fertilizer value but with a consequence of environmental pollution and public health hazard [32]. In the rainy season, ammonia contributes to soil acidification and facilitates algae growth in water bodies [32]. Today, there is growing concern in regulating ammonia emissions from livestock environments worldwide [8, 32]. The concept of litter management has thus far shown drastic reduction of ammonia levels in poultry houses thereby improving birds' health and performance in many places [8, 32]. For instance, ammonia emission is reduced with regular litter change, use of appropriate litter material and amendment facilities, decreased manure moisture and improved indoor conditions [6].

6. Litter amendments

The concept of poultry litter amendments to effectively control ammonia levels involves application of acidifiers, alkaline materials, absorbers, inhibitors, microbial and enzymatic treatments and even dietary manipulations [6, 32, 36].

Acidifiers including alum, sodium bisulphate, ferrous sulphate and phosphoric acid are popular, most effective and widely used poultry litter amenders. They work by creating acidic conditions in litter so that NH_4^+ rather than NH_3^+ is retained, which then facilitates bacteria and enzymatic in activities so that ammonia is not produced in the litter [15, 32, 36]. Alum was reported to reduce NH_3 by 71–92% while phosphoric acid did so by 56–92% [15]. They suppressed NH_3^+ levels below 25 PPM for 3–4 weeks post-application and thus improved in-house air quality in poultry pens [32].

Alkaline materials that include agricultural lime ($CaCO_3$), hydrated or slaked lime Ca $(OH)_2$ or burnt lime (CaO) work by increasing litter alkalinity (pH > 7). This helps to convert more of the HN_4^+ within litter to gaseous NH_3^+ that can be readily lost through provided ventilation so that lower NH_3^+ level is maintained. This practice, however, lowers soluble phosphorus level in litter thus affecting fertilizer value and may subsequently have negative impact on the environment as ammonia levels may later increase significantly when fresh manure is added to such a litter [32, 36].

Absorbers are certain natural clay-type-like zeolites and peats, which are good ammonia absorbers thereby lowering ammonia levels if used in poultry houses [32, 36]. Inhibitors are also used in poultry litter to slowly convert uric acid and urea to ammonia by the process of inhibiting enzymes and microbial activities. Phosphorodiamidate was reported to inhibit urease activity and this reduced the conversion of urea into ammonia [32].

6.1. Microbial and enzymatic treatment of litter

This process utilizes beneficial microbes and enzymes which can convert uric acid and urea rapidly into ammonia which can then be lost out thereby reducing the ammonia levels before chicks are placed in the poultry house. Commercial microbial products like USM-98 or *Yucca schidigera* extract as a natural feed additive were reported to significantly lower ammonia levels, improve bird weights and reduce mortality [32, 36].

6.1.1. Dietary manipulation

This technique involves reducing the nitrogen intake per bird by reducing the crude protein in poultry diet. This works on the concept that ammonia is formed by the breakdown of undigested protein and uric acid in the manure [6]. Therefore, a 1% reduction of CP in poultry diet resulted in 10–22% reduced NH_3 emission in poultry houses [6].

6.1.2. Increased age and weight at slaughter

This process is believed to influence NH_3 emissions because nitrogen excretion per day per bird increases with increasing daily feed intake [10].

7. Processing and utilization of poultry waste

Poultry litter generally contains waste materials including the litter material used, feathers, poultry feed and dead birds, which if properly managed will ensure its beneficial use and will help to prevent adverse effects of improperly disposed litter on the environment and poultry health. Litter is not recommended to be reused when a disease outbreak occurs in a flock [31] because zoonotic pathogens, including *Escherichia coli, Salmonella* spp., *Campylobacter jejuni, Listeria* spp., *Clostridium* spp. and many other viruses, survive in poultry litter for a long time posing health risk to birds [8]. Therefore, litter should be treated to destroy these organisms before land application or before use as feed supplements. Mycotoxins especially aflotoxin B_1 responsible for under-performance in especially broilers have been detected in excess levels in poultry litter [37–40]. Mycotoxins have been known to increase susceptibility of broilers to infectious bursal disease and are reported to also act synergistically with stressor to increase severity of poultry diseases [37]. It is therefore necessary to effectively manage poultry litter before utilization. However, effective management of poultry wastes are normally associated with unbudgeted expenses that are not normally recognized in production budgets, such wastes managed may either be valuable by-products or strictly a net cost on investment.

In many advanced countries, poultry farmers were made to register their operations with appropriate agencies and keep records of poultry wastes so as to help develop an approved poultry waste management plan [41]. Unfortunately, managing poultry waste in most developing countries seems to be impossible and therefore contributing greatly to environmental pollution and disease spread. Below are some practical hints that may render poultry waste easily manageable and ensure environmental safety.

7.1. Converting poultry litter into biofuel

Efforts towards safe disposal of poultry wastes resulted into a technology in the recent pass that converted poultry litter to valuable bio-oil, usable gas and crop fertilizers [42]. For instance, broiler and turkey litters were converted into bio-oils and organic fertilizers and the gas generated in this process was used to operate pyrolysis unit in what seems to be a self-sufficient machine [42]. The machine is made up of a thermochemical unit that destroys pathogenic microorganisms and reduces chances of disease transmission [42]. Furthermore, several electrical generating plants in the UK and recently in the USA utilize poultry and turkey litters as their primary fuel for small-scale electricity generation [42]. Poultry litter is also reported to be used in Ireland as a biomass energy source, and some companies in addition are developing gasification technologies to utilize poultry litter as a fuel for electrical and heating appliances. They are also producing valuable by-products, including activated carbons and fertilizers [43].

7.2. Composting of poultry litter

Composting of poultry litter with dead carcasses is recommended for poultry disease control and an attempt to increase the market value for organic fertilizer generated from poultry litter, which is in high demand in forestry, crop and vegetable farms, homes, lawns and golf courses [44]. Composting is generally a simple natural biological process of converting poultry litter into odourless, stable, consistent and soil-like organic product that is unable to damage crops and surface waters. The process is a slow controlled decomposing or a natural breakdown of organic materials, which utilizes aerobic microorganisms in the poultry litter in the presence of oxygen and moisture to change the chemical and physical nature of poultry litter so that a humus-like material referred to as compost is formed [31, 44]. Composting is believed to reduce litter quantity and weight by 40–80% [31]. Compost has improved air conditioning effects and quality of soils by adding organic matter, nutrients and beneficial microbes thereby increasing soil porosity, density, water and nutrient holding capacity [44]. Compost is thus referred to as an excellent soil amendment [31]. As an example, composted broiler litter has a pH of 5.5–6.5 and is usually weed-free, thus making it a suitable fertilizer for seedlings, shrubs, roses and fruit tries and is also reported to be rich in vitamin B12 [4, 31].

7.3. Storage of poultry litter

The demand for poultry litter is sporadic but highest during the rainy season in many tropical countries thereby requiring temporary holding until the appropriate demanded time. Unfortunately, fresh poultry litter has the highest nitrogen content available for crops making it of greater fertilizer value as at that moment [4]. Notwithstanding proper storage of poultry litter will still ensure its beneficial use as valuable fertilizer and will prevent contamination of surface waters on farms [17]. The most valuable nitrogen in poultry litter is gradually lost to the atmosphere as ammonium over a prolonged period of exposure to the atmosphere [17]. Covered stockpiles of litter: Is a process that involves stockpiles of litter covered using plastic sheets anchored to the earth or other devices to protect against rain and atmospheric losses for timely use. Stockpiles with ground liners: This is another poultry litter storage means involving the use of good plastic sheets as liner to ground or concrete slabs to primarily prevent nutrient leaching to ground water [17]. Permanent storage structures provided with sufficient roofs and concrete floors is the best approach but this is limited by the high risk of spontaneous combustion and fire outbreaks [17].

7.4. Applying poultry litter to crops

Rich soils for efficient crop cultivation were achieved by increased physical fertility due to increased organic matter, increased water holding capacity, increased oxygen diffusion rate that resisted deterioration and disturbance of soil by mining and other industrial activities. Certain considerations are best put in place before litter can be effectively and safely applied on farms [17]. It is logical that poultry manure should not be applied to very steep lands, lands in close proximity to surface waters, drainage ditches and wells for fear of contamination. Likewise application of poultry waste prior to heavy rains is liable to contamination. When applying litter to crops, it is best done at the time of nutrient needs. It should be ensured that litter is applied to as close to planting time as possible or best applied mechanically by incorporating to plants [44]. In this way, ammonia loss due to volatilization and nutrient loss by wind and water erosion are prevented or minimized. Application of litter well ahead of planting time will lead to de-nitrification and leaching [44]. Studies have shown that adequately applied organic manure from poultry litter increased corn yield many folds, but excessive application of poultry manure on the other hand decreased corn yield by a process called 'salt injury phenomena' [20].

7.5. Poultry litter as ruminant feed supplement

Broiler poultry litter contains 25–50% crude protein and 55–60% total digestible nitrogen (TDN) and is also rich in essential minerals. Its nutritional value may even be higher than that of good quality legume hay [45]. Poultry litter has traditionally being used efficiently as a fertilizer; it is now also used as a cost-saving livestock feed supplement for ruminants especially cattle, goats and sheep [46–48]. It is high in urea, a source of nitrogen, which improves the rumen environment making feed more efficiently utilized and the animal better nourished with whatever feed that is made available [46, 48]. Uric acid is a major component of poultry excretions that can be efficiently utilized by rumen microbes for protein production. It is not easily dissolved in the rumen fluid and so the ammonia that is gradually and slowly released is efficiently utilized even more than other non-protein nitrogenous sources [49]. Composted litter is found to be rich in B vitamins especially B12 and can be a good source of this vitamin [31]. The rumen microbe takes about 3 weeks to fully adapt to the utilization of uric acid and so cattle less than 5 months old and sheep and goats less than 3 months old should not be given poultry litter [46]. When poultry litter is processed by an acceptable method, it serves as a very economical and safe source of protein, minerals and energy for many classes of ruminants [46, 48, 49]. Well-processed poultry litter has a total digestible nutrients value similar to average quality hay and this can provide a major portion of the energy to maintain ruminant when fed to them [49]. Poultry litter has been reported to be economically used in ruminant feeding as a forage substitute during drought periods as it also contains high levels of fibre and ash [47, 50]. At higher levels, however, dry poultry litter had depressed growth rate because of its low contents of essential amino acid and excessive amount of calcium in it [49]. It was further shown that poultry manure can replace groundnut cake in the diet of goats without any depressive effects on growth rate and efficiency of feed utilization when used with a good source of carbohydrates such as cassava peel [48]. Crude fibre digestion was reported to be enhanced in rations containing poultry litter and also incorporating poultry litter of up to 25% in the rations of camels did not have any adverse effects [49]. Report of poultry litter as a reliable substitute to cotton seed cake in the diets of suckler cattle was documented [51]. Unfortunately, poultry litter can be contaminated with pathogenic organisms that can cause diseases in other animals but the incidence can be reduced by sun drying [52–55]. For instance, botulism caused by *Clostridium botulinum* has been reported in cattle fed poultry litter [45]. Cupper toxicity was reported in sheep fed on poultry litter when chickens were medicated with cupper as growth promoter in their diets [56]. It may be of scientific concern where poultry feed may contain protein that is prohibited in ruminant feed, such as meat and bone meal particularly where bovine spongiform encephalitis (BSE) had occurred [55] even though this opinion is weak because it seems there has not been any substantial evidence that BSE would survive in chicken intestinal tract. However, wood shavings as the most common litter material in Nigeria are often obtained from wood work industries, and as such may contain sharp or metallic objects which can cause traumatic ventriculitis in poultry and traumatic pericarditis in ruminants [19]. A research concluded that poultry waste intended to be used in compounding rations for cattle should be dried or ensiled and screened for metallic objects to render it safe for use by the animals [19, 57]. Ensiling sorghum forage or molasses with poultry waste had improved crude protein content of the silage almost twofold. Rations formulated with 30% of the concentrate as poultry waste gave about 10 kg of milk/day [55, 58, 59]. However, wet poultry manure should not be fed to livestock and the optimum supplement level for dairy cows is 1–2 kg daily [57].

7.6. Poultry litter as fish feed supplement

Frozen fish is heavily imported into Nigeria [60] and aquaculture is being integrated with livestock and poultry in Nigeria in recent years with a major limitation of formulated fish feeds. Researches have been conducted to determine the effectiveness of feeding cow, pig and poultry manure in variety of fish species [47, 61, 62]. Chickens have short intestines thus excrete about 20% undigested feed and that 10% of feed fed to chickens are wasted to the litter in the process of feeding making available 10–30% total protein content of dry chicken waste. About 1100–1400 Kcal/kg energy and synthesized soluble vitamins are abundant in poultry manure [47, 63]. Some poultry farmers especially in China take this advantage to construct battery cages directly on ponds while others feed poultry manure directly to fish [64], whereas some poultry-fish farmer in Nigeria throw dead poultry into fish ponds so that waste is now recycled into inputs. The nitrogenous waste from poultry litter can efficiently fertilize ponds for growth of plankton as fish food [65]. In fact a report from the United States indicated no difference in terms of growth rate of tilapia raised in poultry litter manure ponds when compared to fish that were fed commercial feed [66].

8. Conclusion

Economic losses as a result of inadequate litter management and utilization are of increasing concern. Selection of litter material is dependent upon cost, availability and quality, which should not be compromised. Research should be continued in search of proper storage conditions of poultry litter and to find the exact inclusion rate of poultry litter to plants and animals. The basic technology of feed compounding using poultry manure and its conversion to useful biofuel can be developed and transferred to farmers.

Author details

Musa I. Waziri* and Bilkisu Y. Kaltungo

*Address all correspondence to: ibwazkalt@yahoo.co.uk

Department of Veterinary Medicine, Veterinary Teaching Hospital, Ahmadu Bello University, Zaria, Nigeria

References

[1] Adene, D.F., 1990. An appraisal of the health management problems of rural poultry stock in Nigeria. In: Proceeding of the International Workshop on Rural Poultry in Africa. Abuja, Nigeria, 989–999.

[2] Moore, P.A., Daniel, T.C, Edwards, D.R., Sharpley, A.N. and Wood, C.W., 2004. Poultry manure management. Journal of Environmental Quality, 36: 60–75.

[3] Durojaiye, A.O., Ahmed, A.S. and Adene, D.F, 1991. Poultry production in Nigeria. Trop. Vet., 44: 37–38.

[4] Embury, I.S., 2004. Alternative litter materials for poultry. Poultry Division of Animal Production, www.agric.nsw.gov.au, NSW Agriculture, 1–5 [accessed, 14 July 2011].

[5] Adewumi, I.K. and Adewumi, A.A., 1996. Managing poultry wastes. In: Workshop on Indigenous Knowledge and Biotechnology. Obafemi Awolowo University, Ile Ife. p. 11.

[6] Meda, A., Hassouna, M., Aubert, C., Robin, P. and Dourmand, J.Y., 2011. Influence of rearing and manure management practices on ammonia and greenhouse gas emissions from poultry houses. World Poultry Science Journal, 67: 441–445.

[7] Asaniyan, E.K., Agbede, A.O. and Laseinde, E.A.O., 2007. Impact assessment of different litter depths on the performance of broiler chicken raised on sand and wood shaving litters. World Journal of Zoology, 2 (2): 67–72.

[8] Musa, I.W., Sa'idu, L., Kaltungo, B.Y., Abubakar, U.B. and Wakawa, A.M., 2012. Poultry litter selection, management and utilization in Nigeria. Asian Journal of Poultry Science, 6: 44–45.

[9] Charles, E.B., 2005. Litter management for confined turkeys. Poultry science and technical guide, No. 41, The North Carolina Agricultural Extension Service Bull, pp. 3–7.

[10] Jacob, J., 2005. Litter materials for small and backyard poultry flocks, University of Kentucky Extension Service, pp. 1–5.

[11] Kelley, T.R., Pancorbo, O.C., Merka, W.C., Thompson, S.A., Carbera, M.I. and Barnhart, H.M., 1995. Bacterial pathogens and indicators in poultry litter during re-utilization. Journal of Applied Poultry Research, 4: 366–373.

[12] Dan Shao, J.O., Jianlu, H.E., Wang, Q., Charry, L., Shi, R.S. and Bing, H.T., 2015. Effects of sawdust thickness on the growth performance, environmental condition and welfare of yellow broilers. Poultry Science, 94: 1–6.

[13] Ndegwa, P.M., Thompson, S.A. and Merka, W.C., 1991. Fractionation of poultry litter for enhanced utilization. Transactions of the American Society of Agricultural Engineers, 34: 992–997.

[14] Wood, C.W., 1992. Broiler litter as a fertilizer. Benefits and environmental concern. In: Proceedings of the National Poultry Waste Management Symposium, Auburn, pp. 305–312.

[15] Moore, P.A., Daniel, T.C., Edwards, D.R. and Miller, D.M., 1996. Evaluation of chemical amendments to reduce ammonia volatilization from poultry litter. Poultry Science, 75: 315–320.

[16] Leo, E., Nathan, S., Morteza, M. and Micheal, D., 2009. The use of poultry litter in row crops. Agriculture and Natural Resources, Cooperative Extension Service, University of Arkansas, pp. 1–3.

[17] Dan, L.C., Casey, W.R. and William, C.M., 2009. Best management practices for storing and applying poultry litter. The University of Georgia Cooperative Extension, pp. 1–12.

[18] Collins, M.S., Gough, R., Alexander, D. and Persons, D., 1989. Virus like particles associated with a wet litter problem in chickens. Veterinary Record, 124 (24): 641.

[19] Musa, I.W., Sai'du, L., Kaltungo, B.Y. and Abdu, P.A., 2011. Common causes of traumatic ventriculitis in layers in Zaria, Nigeria. Veterinary World, 4 (11): 511–514.

[20] Bolan, N.S., Szogi, A.A., Chasavathi, T., Sheshadri, B., Rothrock, M.J. and Panneerselvam, P., 2010. Uses and management of poultry litter. World Poultry Science Journal, 66: 673–698.

[21] Ibrahim, M.A. and Abdu, P.A., 1992. Ethnoveterinary perspective of poultry management health and production among the Hausa/Fulani of rural Nigeria. In: Proceedings of the Scientific Session of the Congress of the Nigerian Veterinary Medical Association, 21–25th October Kaduna, Nigeria, pp. 172–181.

[22] Beri, M.T., 2011. Choice of litter material can decrease salmonella in poultry flocks. World Poultry Journal, 124 (1–2): 71–77.

[23] Shannaway, M.M., 1992. Influence of litter-water holding capacity on broiler weight and carcass quality. Archv-fur-Gefugelkunde., 56 (6): 177–179.

[24] Jesse, L.G., 2004. Alternative litter materials for growing poultry. North Carolina Poultry Industry Newsletter, I (2): 1–5.

[25] Ruszler, P.L. and Carson, J.R., 1968. Physical and biological evaluation of five litter materials. Poultry Science, 41: 246.

[26] Dafwang, I.I., 1990. Rural poultry production in Nigeria; Survey of the production systems and identification of local feed ingredients in rural Nigeria. A report for the Presidential Task Force on Alternative Formulations of Livestock Feeds. pp: 40–47.

[27] Mijinyawa, Y. and Dlamini, B.J., 2006. Livestock and poultry waste management in Swaziland. Livestock Research for Rural Development, 18 (6): 1–12.

[28] Terzich, M., Quarles, C., Goodwin, M.A. and Brown, J., 1998. Effect of poultry litter management on the development of respiratory lesions in broilers. Avian Pathology, 27: 566–569.

[29] Ezeokoli, C.D., Umoh, J.U., Adesiyun, A.A. and Abdu, P.A., 1984. Poultry production in Nigeria. The Bulletin of Animal Health and Production in Africa, 10 (32): 253–257.

[30] Anonymous, 1990. General note on stock management (litter management). Poultry Management Guide, 4–5.

[31] Anne, F., 2007. Poultry house management for alternative production. Poultry International. www.poultryinternational.digital.com. 1–10. [accessed 5 July 2011].

[32] Sanjay, S., Philip, W. and James, P., 2006. Poultry litter amendments. North Carolina Cooperative Extension Service Bull., pp. 1–6.

[33] Ritz, C.W, Fairchild, B.D. and Lacy, M.P., 2004. Implications of ammonia production and emission from poultry facilities: a review. Journal of Applied Poultry Research, 13: 684–692.

[34] Wheeler, E.F., Casey, K.D., Zajaczkowski, J.L., Topper, P., Gates, R.S., Xin, H. and LIANG, H., 2004. Seasonal ammonia emission variation among twelve U.S broiler houses. ASAE paper 044105. American Society of Agricultural and Engineers, pp. 182–187.

[35] Xin, H., Tanaka, A., Wang, T., Gates, R.S., Wheeler, E.F., Casey, K.D., Herber, A.J., Ni, J. and Lim, T., 2002. A portable system for continuous ammonia measurement in the field. American Society of Agricultural Engineers., 125.

[36] John, P.B. and Joseph, B.H., 2001. Litter treatment for poultry. Alabama Cooperate Extension System. ANR: 1199, 1–3.

[37] Westiake, K., 1985. The incidence of mycotoxins in litter, feed and livers of chickens in Natal. South African Journal of Animal Science, 15 (4): 175–177.

[38] Manafi, M., Mohan, K. and Noor Ali, M., 2011. Effect of ochratoxin A on coccidiosis-challenged broiler chicks. World Mycotoxin Journal, 4(2): 177–181.

[39] Viegas, C., Carolino, E., Malta-Vacas, J., Sabino, R., Viegas, S. and Verissimo, C., 2012. Fungal contamination of poultry litter: a public health problem. Journal of Toxicology and Environmental Health, 75 (22–23): 1341–1350. DOI: 10.1080/15287394.2012.721165

[40] Manafi, M., Pirany, N., Noor Ali, M., Hedayati, M., Khalaji, S. and Yari, M., 2015. Experimental pathology of T-2toxicosis and mycoplasma infection on performance and hepatic functions of broiler chickens.

[41] Damona, D., Brian, F. and Joshua, P., 2009. Broiler production considerations for potential growers. Oklahoma Cooperative Extension fact sheet. http://www.osuextra.com [accessed 25April 2011].

[42] Anonymous, 2007. Converting poultry litter in to bio-oil. Poultry International. www.poultryinternational.digital.com. pp 1–3 [accessed 12 June 2011].

[43] Singh, R.P., Dhania, G., Sharma, A. and Jaiwal, P.K., 2007. Biotechnological approaches to improve phytoremediation efficiency for environment contaminants. In: Environmental bioremediation technologies, Singh, S.N., Tripahti, R.D. (Eds). Springer, pp. 223–258.

[44] Forbes, W., 2006. On-farm composting of poultry litter. Agricultural Extension Service Unit, the University of Tennessee Institute of Agriculture, pp. 1–9.

[45] Fontenot, P.H. and Hancock, W.J., 2001. Utilization of poultry litter as feed for beef cattle. A Paper Presented at FDA Public Hearing on Animal Feeding Regulation 'Animal Proteins Prohibited in Ruminant Feed' Kansas City, MO, October 30th, Title 21, Part 589–2000.

[46] Murthy, K.S., Reddy, M.R. and Reddy, G.V.N., 1995. Utilization of cage layer droppings and poultry litter as feed supplements for lambs and kids. Small Ruminant Research, 16 (3): 221–225.

[47] Anonymous, 2006. Better utilization of locally available feed resources. http://www.fao.org/ag/againfo/themes/documents [accessed 13 July 2011].

[48] Adegbola, A.A., Smith, O.B. and Okeudo, N.J., 2010. Response of West African dwarf sheep fed cassava peel and poultry manure based diets, FAO Corporate Document Repository produced by ILRI. pp. 1–8.

[49] Abdel-Baset, N.S. and Abbas, S.F., 2010. Study on the use of dried poultry litter in the camel's ration. Veterinary Research Forum, 2 (1): 65–71.

[50] Chauhan, T.R., 1993. Nutritional evaluation of urea/poultry litter enriched wheat straw based rations to adult buffaloes. Buffalo Journal, 9 (3): 187–194.

[51] Bayemi, P.H., Tiabon, E.B., Nguemdjom, A., Kamga, P., Mbanya, J., Ndi, C., Nfi, A. and Mangeli, E., 2001. Effect of replacing cotton seed cake with poultry droppings on weight gain of growing cattle at Bambul, Cameroon. Tropical Animal Health and Production, 33 (1): 49–56.

[52] Khattab, H.M., El-Sayed, H.M. and Elwan, K.M., 1995. Poultry litter incorporation in the fattening ration of male buffalo calves. In: Proceedings of 5th Scientific Conference on Animal Nutrition. 1: 145–151.

[53] Gabr, A.A., El-Ayek, M.Y and Mehrez, A.Z, 2003. Effect of long term feeding of ration containing dried poultry litter on digestibility and growing lamb performance. Journal of Agricultural Science, 18: 3437–3438.

[54] Adene, D.F, 1997. Diseases of poultry in Nigeria. An overview of problems and solutions. Tropical Veterinarian, 15: 103–110.

[55] Anonymous, 2011. Poultry litter from Wikipedia, the free encyclopedia. www.extension.orgeXtension.community [accessed 12 July 2011].

[56] Khattab, H.M., El-sayed, H.M. and El-man, K.M., 1995. Poultry litter incorporation in the fattening ration of male buffalo calves. Proc. Sci. Conf. Nurt., 1: 145–151.

[57] Abdul, S.B., Yashim, S.M. and Jocthan, G.E., 2008. Effects of supplementing sorghum stover with poultry litter on performance of wandara cattle. American-Eurasian Journal of Agronomy, 1 (1): 16–18.

[58] Odhuba, E.K., 2006. Towards efficient utilization of poultry waste by ruminants. Gttp://www.ilri.org/infoserve/webpub/fulldocs/15490E/x5490E09 [accessed 2 June 2010].

[59] Owen, O.J., Ngodigba, E.M. and Amakiri, A.O., 2008. Proximate comparison of heat treated poultry litter (layer). International Journal of Poultry Science, 1–14.

[60] Solarin, B.B., 1992. Aspect of the fishing industry and an overview of the artisanal reefs and fish aggregating devices for increasing fisheries output and viability in Nigeria. In: Proceedings of 10th Annual National Conference of Fisheries (FISCON), 25–30 November, Abeokuta, Nigeria, pp. 89–94.

[61] Smitherman, R.O., Shelton, S.W. and Grover, J.H. (Eds.). The American Fisheries Society, Aubum Alabama. pp: 43–54.

[62] Campos, E. and Sampao, I., 1976. Swine feces recycling in carp feeding. Arg. ESC. Vet. U.F.M., 28: 147–152.

[63] Tuleun, C.D., 1992. The utilization of heat-treated poultry manure in chicks diets. Paper presentation at the 1st Annual Conference of the National Society of Animal Production, 26–29 September, Abuja, Nigeria.

[64] Rangayya, V., 1977. Poultry cum fish farm. ("The Hindu"). Asian Livestock, 3 (3): 1–5.

[65] Adewumi, A.A., Adewumi, I.K. and Olaleye, V.F., 2011. Livestock waste menace: Fish wealth-solution. African Journal of Environmental Science and Technology, 5 (3): 149–154.

[66] Collins, W.J. and Smitherman, I., 1978. Production of tilapia hybrids with cattle manure or a commercial diet. In: Symposium on culture of exotic fishes.

Permissions

All chapters in this book were first published in PS, by InTech Open; hereby published with permission under the Creative Commons Attribution License or equivalent. Every chapter published in this book has been scrutinized by our experts. Their significance has been extensively debated. The topics covered herein carry significant findings which will fuel the growth of the discipline. They may even be implemented as practical applications or may be referred to as a beginning point for another development.

The contributors of this book come from diverse backgrounds, making this book a truly international effort. This book will bring forth new frontiers with its revolutionizing research information and detailed analysis of the nascent developments around the world.

We would like to thank all the contributing authors for lending their expertise to make the book truly unique. They have played a crucial role in the development of this book. Without their invaluable contributions this book wouldn't have been possible. They have made vital efforts to compile up to date information on the varied aspects of this subject to make this book a valuable addition to the collection of many professionals and students.

This book was conceptualized with the vision of imparting up-to-date information and advanced data in this field. To ensure the same, a matchless editorial board was set up. Every individual on the board went through rigorous rounds of assessment to prove their worth. After which they invested a large part of their time researching and compiling the most relevant data for our readers.

The editorial board has been involved in producing this book since its inception. They have spent rigorous hours researching and exploring the diverse topics which have resulted in the successful publishing of this book. They have passed on their knowledge of decades through this book. To expedite this challenging task, the publisher supported the team at every step. A small team of assistant editors was also appointed to further simplify the editing procedure and attain best results for the readers.

Apart from the editorial board, the designing team has also invested a significant amount of their time in understanding the subject and creating the most relevant covers. They scrutinized every image to scout for the most suitable representation of the subject and create an appropriate cover for the book.

The publishing team has been an ardent support to the editorial, designing and production team. Their endless efforts to recruit the best for this project, has resulted in the accomplishment of this book. They are a veteran in the field of academics and their pool of knowledge is as vast as their experience in printing. Their expertise and guidance has proved useful at every step. Their uncompromising quality standards have made this book an exceptional effort. Their encouragement from time to time has been an inspiration for everyone.

The publisher and the editorial board hope that this book will prove to be a valuable piece of knowledge for researchers, students, practitioners and scholars across the globe.

List of Contributors

Farai Catherine Muchadeyi
Agricultural Research Council, Biotechnology Platform, Pretoria, South Africa

Ufuk Tansel Sireli
Department of Food Hygiene and Control, Faculty of Veterinary Medicine, Ankara University, Ankara, Turkey

Ayhan Filazi and Begum Yurdakok-Dikmen
Department of Pharmacology and Toxicology, Faculty of Veterinary Medicine, Ankara University, Ankara, Turkey

Ozgur Kuzukiran
Veterinary Control Central Research Institute, Ankara, Turkey

Ufuk Tansel Sireli
Department of Food Hygiene and Control, Faculty of Veterinary Medicine, Ankara University, Ankara, Turkey

Dorota Witkowska and Janina Sowińska
Department of Animal and Environmental Hygiene, Faculty of Animal Bioengineering, University of Warmia and Mazury in Olsztyn, Olsztyn, Poland

Michel Federighi
Oniris, Food Quality and Hygiene, Nantes, France

Daise A. Rossi, Roberta T. Melo, Eliane P. Mendonça and Guilherme P. Monteiro
Laboratory of Applied Animal Biotechnology, Faculty of Veterinary Medicine, Federal University of Uberlândia, Uberlândia, Brazil

Anicke Brandt-Kjelsen and Brit Salbu
Department of Environmental Sciences, Norwegian University of Life Sciences (NMBU), Ås, Norway

Anna Haug
Department of Animal and Aquacultural Sciences, Norwegian University of Life Sciences (NMBU), Ås, Norway

Joanna Szpunar
Laboratorie de Chimie Analytique Bio-inorganique et Environement, CNRS UMR 5254, Pau, France

Hong Li, Zhuanjian Li and Xiaojun Liu
Henan Agricultural University, Zhengzhou, Henan, China

Herbert K. Dei
Department of Animal Science, Faculty of Agriculture, University for Development Studies, Tamale, Ghana

Daria Murawska
Department of Commodity Science and Animal Improvement, Faculty of Animal Bioengineering, University of Warmia and Mazury in Olsztyn, Olsztyn, Poland

Musa I. Waziri and Bilkisu Y. Kaltungo
Department of Veterinary Medicine, Veterinary Teaching Hospital, Ahmadu Bello University, Zaria, Nigeria

Index

A
Abdominal Fat, 188-189, 192, 194, 198-201 204
Adhesion, 100-103, 106-107, 110-112, 115-116
Aflatoxin, 41-42, 52-54, 56-59, 67, 72, 77, 80
Amino Acids, 122, 125-129, 133-134, 138 149, 157, 163-166, 169-170, 173, 177, 182
Animal Feed, 53, 121, 124, 127, 139, 141 169-170, 187
Antibiotic, 31-32, 34, 39, 51, 105, 114-115, 138
Antimicrobial, 22, 35, 88, 92, 98-99, 104, 108 112, 114, 119-120
Arsenic, 24, 137, 140, 207

B
Bentonite, 50, 58, 60, 71-72, 80
Bioaerosol, 62, 64, 68
Bioavailability, 56, 108, 124, 126, 130, 134 137, 140-141, 167
Biofilm, 100-120
Biofilm Formation, 100, 102-103, 105-112 114-119
Biosecurity, 4, 62, 86-87, 89, 93, 96, 105-106
Body Weight, 46-47, 54, 72, 174, 188-202
Breed, 4, 6-8, 11-13, 17, 191, 203
Breeding, 2, 5, 7, 11-14, 16-19, 30, 43, 78, 80 82, 86-87, 91-92, 158, 168, 183, 185, 189 194, 204
Broiler, 23, 35, 43-44, 46, 51, 53-59, 62, 67, 70 73-74, 76-81, 86-88, 91-98, 101, 116, 138 141-142, 183, 186-192, 194-195, 197-198 200-201, 203, 205, 209-210, 215-217, 219-222
Broiler House, 62, 74, 76, 79, 87

C
Campylobacteriosis, 82-87, 96, 101
Carcinogenic, 20, 22, 25, 28-29, 42-43, 45, 53 55
Chickens, 1-19, 35, 43, 48, 54, 56-58, 73, 75 77-78, 80-81, 86-92, 94, 96-99, 105, 121, 124 127-128, 130-136, 141-143, 145, 169, 174 176-178, 187-192, 194-195, 197-198 200-201, 203, 205, 209, 218, 220, 222

D
Detoxification, 42-43, 49-53, 58-59
Dietary Treatment, 48, 128, 130-131, 133, 136

E
Edible Component, 189, 200-201
Egg Production, 3, 5, 7, 37, 40, 42, 44, 46, 48 51, 123, 144, 177-178, 189
Endotoxin, 68, 76-77

Enzymes, 40, 42-43, 49-51, 53, 111-112, 119 144, 146-147, 151, 214
Essential Oil, 60, 70, 75-76, 79

F
Fatty Acids, 30, 39, 43, 61, 69, 88, 97, 112 143, 145-146, 148-149, 153, 162, 198
Feed Additive, 23-24, 50-51, 97, 123, 139, 214
Feed Supplement, 57, 206-207, 217-218
Fumonisin, 45, 55-57, 59
Fungi, 22, 40-41, 49-50, 52, 60-64, 66-68, 74 139, 146, 171

G
Geese, 70, 188-201, 203-204
Gene Expression, 11, 17, 35, 43-44, 54, 118 127, 129, 134, 136, 142, 144-146, 148, 150 153
Genetic Adaptation, 1, 4-8, 10
Genetic Diversity, 1-3, 6-14, 16-17, 43, 204
Genotype, 17, 204-205
Giblets, 24, 35, 188-189, 192, 196-197, 200-201
Gizzard, 44, 188-189, 192, 196-197

H
Halloysite, 60, 71-72, 76, 79
Hazard, 49, 58, 68, 82, 86, 206, 214
Hepatic Lipid, 143-146, 148-150, 153, 156

L
Lead, 23-32, 35, 43, 51, 68-69, 75, 108, 123 197, 209, 211-213, 217
Lean Meat, 189, 192-195, 203
Lipid, 27, 58, 142-156, 162-163
Lipid Metabolism, 143-150, 152-156
Lipoprotein, 144, 147, 149, 151-152, 154-155
Litter Management, 207, 211-214, 218 220-221
Litter Material, 207-215, 218, 221
Lysine, 134, 157, 163-166, 168-174, 176-179 181-187

M
Metabolism, 42-45, 53-55, 103, 106-107, 117 121, 124-126, 135-136, 139-141, 143-156
Metabolite, 43-44, 51
Methionine, 122, 132, 134, 164-165, 173-174 176-177, 187
Moisture Content, 71-74, 168, 170-171, 185 212
Mortality, 4, 23, 40, 44, 54, 72, 84, 94, 129 195, 209, 214
Motility, 103, 106-107, 109, 117

Muscle Tissue, 26-27, 35, 134, 188-189 191-195, 200-202

Mycotoxin, 40-41, 46-47, 49-51, 53-56, 58-59 79-80, 222

O

Ochratoxin, 41, 43-45, 53-54, 57-58, 67, 72, 77 79-80, 222

Offal, 24-25, 189, 192, 194, 198-199, 201-202

P

Pekin Ducks, 188-189, 191, 193-197, 199-200 203

Phage, 98, 112, 120

Phthalates, 20, 22, 29-30, 38

Poultry House, 24, 27, 62, 73, 77, 87, 212-214 221

Poultry Litter, 35, 72, 74, 79, 206-209, 211 214-223

Poultry Manure, 77, 79, 206-207, 217-219 222-223

Poultry Meat, 20-25, 31, 33, 41, 73, 87, 95-96 188-190, 199, 201

Poultry Production, 15-17, 21-22, 28, 39, 41 56, 69-70, 72, 93, 104-106, 157, 159-160 178-179, 181, 198, 201, 219, 221

Poultry Species, 3, 41, 70, 188-189, 191-192 198-200

Poultry Waste, 206-207, 215, 217-218 220-221, 223

Protein, 5, 8, 17, 42, 44, 46-48, 57, 72, 89-91 97, 101, 106-108, 121, 123, 125, 129 133-136, 144, 146, 152-166, 168-187, 206 215, 217-218

Protein Content, 123, 157-158, 161, 163, 165 168-169, 171, 179, 183, 186, 218

Public Health, 24-25, 34, 40-43, 89, 94-95 100-101, 113, 206, 214, 222

Q

Quality Protein Maize, 157, 171, 179-181 183, 185-187

S

Selenite, 121-123, 125-127, 132-136, 139, 141

Selenoprotein, 125-126, 132, 140

Speciation, 124-125, 127-128, 133, 136 139-142

Strain, 51, 58, 92, 99, 105, 111

Subcutaneous Fat, 188-189, 191, 194-196 200-201

T

Tissue Component, 194

Turkeys, 23, 42-43, 45-46, 53, 55, 63-64, 74-75 130, 138, 141, 188-195, 197-205, 209, 220

V

Ventilation, 49, 62-64, 67-71, 74, 76, 79, 105 207, 211-212, 214

Veterinary Drug, 23, 33-35, 39

Vitamin, 137, 141-142, 149, 163, 166-168 216-217

W

Wood Shaving, 63, 208-209, 219

CPSIA information can be obtained
at www.ICGtesting.com
Printed in the USA
BVHW011244270822
645617BV00003B/122